NANOTECHNOLOGY FOR CIVIL INFRASTRUCTURE

NANOTECHNOLOGY FOR CIVIL INFRASTRUCTURE

Innovation and Eco-efficiency of Nanostructured Cement-Based Materials

Edited by

KAMAL H. KHAYAT

Department of Civil, Architectural and Environmental Engineering, Missouri University of Science and Technology, Rolla

WEINA MENG

Department of Civil, Environmental and Ocean Engineering, Stevens Institute of Technology, Hoboken

ELSEVIER

Elsevier
Radarweg 29, PO Box 211, 1000 AE Amsterdam, Netherlands
The Boulevard, Langford Lane, Kidlington, Oxford OX5 1GB, United Kingdom
50 Hampshire Street, 5th Floor, Cambridge, MA 02139, United States

ISBN: 978-0-12-817832-4

For information on all Elsevier publications
visit our website at https://www.elsevier.com/books-and-journals

Publisher: Matthew Deans
Acquisitions Editor: Ana Claudia Garcia
Editorial Project Manager: Fernanda A. Oliveira
Production Project Manager: Fizza Fathima
Cover Designer: Greg Harris

Typeset by STRAIVE, India

Contents

Contributors

Nima Farzadnia
Department of Civil, Geological, and Environmental Engineering, University of Alaska Fairbanks, Fairbanks, AK, United States

Reed T. Heintzkill
Department of Materials Science & Engineering, College of Engineering and Applied Science, University of Wisconsin–Milwaukee, Milwaukee, WI, United States

Qingxu Jin
Department of Civil and Environmental Engineering, Michigan State University, East Lansing, MI, United States

Kamal H. Khayat
Center for Infrastructure Engineering Studies, Department of Civil, Architectural and Environmental Engineering, Missouri University of Science and Technology, Rolla, MO, United States

Marina I. Kozhukhova
Department of Civil & Environmental Engineering, College of Engineering and Applied Science, University of Wisconsin–Milwaukee; Ramboll, Milwaukee, WI, United States

Wu-Jian Long
Guangdong Provincial Key Laboratory of Durability for Marine Civil Engineering, Shenzhen Durability Centre for Civil Engineering, College of Civil and Transportation Engineering, Shenzhen University, Shenzhen, Guangdong, PR China

Hongyan Ma
Department of Civil, Architectural and Environmental Engineering, Missouri University of Science and Technology, Rolla, MO, United States

Weina Meng
Department of Civil, Environmental and Ocean Engineering, Stevens Institute of Technology, Hoboken, NJ, United States

Caijun Shi
Key Laboratory for Green & Advanced Civil Engineering Materials and Application Technology of Hunan Province, International Innovation Center for Green & Advanced Civil Engineering Materials of Hunan Province, College of Civil Engineering, Hunan University, Changsha, China; Department of Civil Engineering, The University of British Columbia, Vancouver, BC, Canada

Konstantin Sobolev
Department of Civil & Environmental Engineering; Department of Materials Science & Engineering, College of Engineering and Applied Science, University of Wisconsin–Milwaukee, Milwaukee, WI, United States

Zemei Wu
College of Civil Engineering, Hunan University, Changsha, People's Republic of China; Center for Infrastructure Engineering Studies, Department of Civil, Architectural and Environmental Engineering, Missouri University of Science and Technology, Rolla, MO, United States

Foreword

The book *Nanotechnology for Civil Infrastructure* edited by Kamal Khayat and Weina Meng is very timely and provides comprehensive and well-documented information on the applications of nanotechnology to cement-based materials. In the past couple of decades, there has been increasing interest in incorporating nanomaterials into concrete. The motivation behind these explorations is whether one can significantly enhance the strength, durability, and sustainability of concrete construction. Concrete is one of the most widely used man-made materials. The current world production of concrete is about 2 tons per living human being. With the trend toward increasing urbanization in developing countries and with the need for rehabilitation of infrastructure in developed countries, the demand for concrete is likely to be acute. However, the manufacture of concrete requires heavy use of depleting natural resources and produces a substantial amount of greenhouse gases, contributing to global warming. Industry and academia are working very hard to develop new approaches and innovative materials to make concrete more sustainable. Nanotechnology will certainly play a key role in this new development. This book provides a valuable guide to the challenges and opportunities in applying nanotechnology to concrete construction.

The book is divided into eight chapters. Each chapter is authored by individuals who have conducted considerable research on the subject of the chapter. The nanoparticles that the book addresses include primarily nanosilica, nanoalumina, nanotitanium dioxide, carbon nanotubes, graphene, and graphene oxides. The key to the successful incorporation of nanoparticles is their homogeneous dispersion. The issues of dispersion and potential modification of the surfaces of nanoparticles to aid uniform dispersion are well detailed in the book.

The book describes how various nanoparticles influence the hydration reaction, microstructure, mechanical properties, and durability of cementitious composites. The book discusses the physical and chemical mechanisms that are responsible for the enhanced performance of concrete reinforced with nanoparticles.

In addition to conventional Portland cement, concrete may contain supplementary cementitious materials (such as fly ash) and alkali-activated slag or fly ash. The book devotes a chapter to the application of nanomaterials to geopolymer systems.

One of the successful applications of nanoparticles is the so-called self-cleaning cement-based material. These are based on the photocatalytic property of titanium dioxide. The last chapter of the book details the application of these materials.

Professor Kamal Khayat and his group have been extensively involved in many aspects of nanomaterials and their applications. Their experience and knowledge make this book a valuable source for researchers as well as practitioners.

I congratulate the editors, Dr. Kamal Khayat and Dr. Weina Meng, for compiling and synthesizing a substantial amount of information.

Surendra P. Shah[a,b]

[a]Presidential Distinguished Professor, Director, Center for Advanced Construction Materials, University of Texas at Arlington, Arlington, TX, United States

[b]Walter P. Murphy Professor (Emeritus), Northwestern University, Evanston, IL, United States

CHAPTER 1

Introduction

Kamal H. Khayat[a] and Weina Meng[b]
[a]Center for Infrastructure Engineering Studies, Department of Civil, Architectural and Environmental Engineering, Missouri University of Science and Technology, Rolla, MO, United States
[b]Department of Civil, Environmental and Ocean Engineering, Stevens Institute of Technology, Hoboken, NJ, United States

Cement-based material is the most used construction material worldwide. It consists of aggregates and sometimes fibers bonded together with amorphous form, nanometer to micrometer crystals (i.e., cementitious materials hydration products). The nanotechnologies and related nanomaterials have been used in cement-based materials to modify the form and density of the crystals since the 1980s [1,2]. The term nanomaterials coined in the early 1980s wherein the majority size lies in the range of 1–500 nm, with a great specific surface area, emphasize being given to metallic, nonmetallic, carbon-based, and mineral nanoparticles [3]. In recent years, with the extending research of materials science and technology, the costs of the nanomaterials have decreased, and the availability of nanomaterials has increased for various levels of applications in civil construction [4]. Examples of the use of nanomaterials in concrete construction include the Jubilee Church in Rome constructed in 2003 and the Gärtner-platzbrücke Bridge in Germany built in 2007.

The nano-sized particles can result in a dramatic improvement in some of the key concrete properties and add new features in cement-based materials compared to conventional grain-sized materials of the same chemical composition. By filling the pores between large cementitious particles and providing more nucleation sites ofhydration products, when properly dispersed nanomaterials, the structure of hydration gel providing a dense and solid hydration structure which significantly increase in the workability and strength of cement-based materials can be achieved [5]. Nanoparticles can greatly refine the interfacial zones between cementitious systems and aggregates/fibers and produce greater and higher resistance to cracking of such transition zones. The use of nanomaterials can manipulate and alter the cement matrix system by providing a new nanoscale structure. Common discrepancies in concrete microstructure, such as microvoids, capillary porosity, and deterioration due

Nanotechnology for Civil Infrastructure
https://doi.org/10.1016/B978-0-12-817832-4.00005-5

to alkali-silicate reaction, freeze-and-thaw, corrosion, or sulfate attack can be greatly reduced. In addition, through the combination of the filler effect and additional bridge effect, new nano-modified carbon-based materials with greater durability and enhanced tensile/flexural performance can be developed [5]. Further, the use of nanomaterials can provide useful functional properties to the cement-based materials, such as self-cleaning [6], self-sensing [7], self-healing [8], and temperature management [9].

This book explores recent innovations in cement-based materials developed through nanotechnology. Of special interest is highlighting the advantages that can be associated with the use of nanomaterials that can enhance performance compared with conventional products and potentially reduce resources and energy consumption throughout the life cycle of cement-based materials used in the construction and repair industry.

The first part of the book introduces the nanomaterials that are typically employed for cement-based materials and their effect on performance and eco-efficiency in the concrete industry.

Chapter 2 reviews available surface modification methods for inorganic nanoparticles, namely, nanosilica, nanotitania, and nanoalumina, as well as carbon nanofilaments, such as carbon nanofibers, carbon nanotubes, and graphene platelets. Finally, the chapter discusses the suitable modification techniques of nanoparticles for use in cement-based materials and their effects on the performance of cement-based materials. Although limited studies are available in this field, modification of nanomaterials for better dispersion is the first step to be considered, which is the key to advance multi-functionalities in cement-based materials.

Chapter 3 provides an overview of contemporary ongoing research, postulated future developments, and general zones of innovation in which nanotechnology may be implemented into cement and concrete materials to enhance ecologic impacts and maximize long-term sustainability.

Part two of the book discusses basic principles related to the chemical and physical phenomena of 0-D, 1-D, and 2-D reinforced effects occurring at the nanoscale, which is responsible for the final performances of cementitious materials. Such principles are key to optimize the properties of cement-based materials, including rheological properties, mechanical properties, transport properties, durability, and eco-efficiency.

Chapter 4 discusses the common 0-D nanomaterials employed in cement-based materials, which include nano-SiO_2, nano-$CaCO_3$, nano-Fe_2O_3, nano-Fe_3O_4, nano-TiO_2, nano-CuO, and nano-ZiO_2. The dispersion and mixing procedure, mechanisms underlying performance enhancement, and how those materials affect the fresh and hardened properties of cement-based materials are detailed.

Chapter 5 reviews recent research on the design, performance, and mechanism of cement-based materials with commonly used 1-D nanomaterials, including carbon nanotubes and carbon nanofibers.

Chapter 6 comprehensively reviews the research progress on the design, performance, and mechanism of cement-based materials with 2-D nanomaterials, mainly graphene nanosheets and their derivatives.

In part three, the book reviews some special applications of nanomaterials on cement-based materials.

Chapter 7 discusses the application of nanomaterials used in geopolymer systems. The effects of the widely used nanomaterials on the reaction kinetics, chemical characterization of reaction products, pore structure, mechanical properties, as well as durability of geopolymers are comprehensively reviewed. In this chapter, the term "geopolymer" is used as an interchangeable term for "alkali-activated materials."

Chapter 8 introduces the application of nanomaterials on "self-cleaning" cement-based materials, also called nano-TiO_2-engineered photocatalytic cementitious materials. Firstly, the two main types of TiO_2-engineered cementitious materials and self-cleaning mechanisms are introduced. Afterward, the effects of TiO_2 on the cement for each type, including hydration process, microstructure, strength development, surface hydrophilicity, and bond strength between coating and cementitious substrates are discussed. Finally, the application of methylene blue (MB) photodegradation is presented as an example of the self-cleaning of nanomaterials for concrete structures.

References

[1] O.S. Olafusi, E.R. Sadiku, J. Snyman, J.M. Ndambuki, W.K. Kupolati, Application of nanotechnology in concrete and supplementary cementitious materials: a review for sustainable construction, SN Appl. Sci. 1 (6) (2019) 1–8.
[2] Z. Zhao, T. Qi, W. Zhou, D. Hui, C. Xiao, J. Qi, Z. Zheng, Z. Zhao, A review on the properties, reinforcing effects, and commercialization of nanomaterials for cement-based materials, Nanotechnol. Rev. 9 (1) (2020) 303–322.
[3] P. Balaguru, C. Ken, Nanotechnology and concrete: research opportunities, in: Proceedings of the ACI Session on Nanotechnology of Concrete. Recent Developments and Future Perspectives, 2006, pp. 15–28.
[4] J. Silvestre, N. Silvestre, J. De Brito, Review on concrete nanotechnology, Eur. J. Environ. Civ. Eng. 20 (4) (2015) 455–485.

[5] W. Meng, K.H. Khayat, Mechanical properties of ultra-high-performance concrete enhanced with graphite nanoplatelets and carbon nanofibers, Compos. Part B 107 (2016) 113–122.
[6] S.N. Zailan, N. Mahmed, M. Abdullah, A.V. Sandu, N.F. Shahedan, Review on characterization and mechanical performance of self-cleaning concrete, in: MATEC Web of Conferences 97: 01022, EDP Sciences, 2017.
[7] B. Han, S. Ding, X. Yu, Intrinsic self-sensing concrete and structures: a review, Measurement 59 (2015) 110–128.
[8] N. Shaheen, R.A. Khushnood, W. Khaliq, H. Murtaza, R. Iqbal, M.H. Khan, Synthesis and characterization of bio-immobilized nano/micro inert and reactive additives for feasibility investigation in self-healing concrete, Constr. Build. Mater. 226 (2019) 492–506.
[9] A. Jurizat, S. Wonorahardjo, A review on the application of phase change material for indoor temperature management in tropical area, IOP Conf. Ser.: Earth Environ. Sci. Vol. 152 (1) (2018) 012022. IOP Publishing.

CHAPTER 2

Modification of nanomaterials for nanostructured cement-based materials

Nima Farzadnia[a] and Kamal H. Khayat[b]
[a]Department of Civil, Geological, and Environmental Engineering, University of Alaska Fairbanks, Fairbanks, AK, United States
[b]Center for Infrastructure Engineering Studies, Department of Civil, Architectural and Environmental Engineering, Missouri University of Science and Technology, Rolla, MO, United States

2.1 Introduction

Nanomaterials are materials with at least one dimension on the nanoscale (100 nm or less) that can occur either naturally [1] or synthetically [2]. The term nanomaterial was first coined by an American scientist, Richard Feynman, at the American Physics Society in 1959 [3]. He established the conceptual foundations of nanotechnology by saying, "there is plenty of room at the bottom." Today, the production of engineered nanomaterials with unique physical and chemical properties is rapidly expanding for different applications, such as bioengineering, medicine, the automobile industry, energy, and construction.

There are two approaches for nanomaterial fabrication, namely, top-down and bottom-up [4]. Both are very important in modern industry. The top-down approach initiates the fabrication of nanomaterials from the larger-scale structure of the material and reduces the size of the nanoscale. Ball milling and application of severe plastic deformation are typical examples of this method. The main problem of this method is the surface structure imperfection and possible crystallographic damage caused during the process, which lead to the slow and expensive production of nanoparticles. The bottom-up approach refers to the buildup of nanomaterials from the bottom, starting with an atom, molecule, or building cluster. This method is less expensive compared to the top-bottom process.

In general, nanomaterials received significant attention because of their unique properties. Nanomaterials have a much greater surface area-to-volume ratio than their bulk crystal, leading to greater chemical reactivity. In recent years, an increasing trend to use nanotechnology in different fields

Nanotechnology for Civil Infrastructure
https://doi.org/10.1016/B978-0-12-817832-4.00008-0

has encouraged research to apply physical and chemical modification of nanoparticles to maximize their unique benefits in various applications. This chapter reviews available surface modification methods for inorganic nanoparticles, namely, nanosilica (NS), nanotitania (NT), and nanoalumina (NA), as well as carbon nanofilaments, such as carbon nanofibers (CNFs), carbon nanotubes, and graphene platelets. Finally, the chapter discusses the most recent modification techniques of nanoparticles for cement-based materials and their effects on their performance. Although limited studies are available in this field, modification of nanomaterials can lead to their advanced functionality in cement-based materials.

2.2 Nanomaterials in cement-based materials; Benefits and challenges

Nanomaterials are used in cement-based materials to improve their engineering properties and functional performance. They can also contribute to the sustainability of cement-based materials by enhancing the durability and reducing the amount of required cement to cast cement-based materials. At the fresh state, nanomaterials improve the stability of the mixtures by enhancing the segregation/bleeding resistance [5]. This is achieved by increasing the mixture's viscosity at the fresh state due to the high surface energy of the nanoparticles [6]. During the hydration process, nanoparticles can act as nuclei for the hydrated products and help better distribute the calcium-silicate-hydrate (C—S—H) gel in the matrix. This phenomenon can lead to a more homogeneous cement matrix [7]. Besides, nanomaterials can accelerate the dissolution of C_3S and C_2S through their high surface energy and accelerate C—S—H gel formation [8]. Nanoparticles with high pozzolanic performance, such as NS, can react with calcium hydroxide and form additional (secondary) C—S—H gel, which reduces calcium leaching and improves the microstructure of the cement matrix.

Furthermore, the nanofilling effect of particles densifies the structure of the cement matrix at the hardened stage by filling the interlayer pores between C—S—H and contributing to the concrete strength [9]. They can also improve the bond between aggregates and bulk cement paste by enhancing the structure of the contact zone due to their high surface energy [10]. Nanoparticles, especially with a high aspect ratio, may have interlocking effects between failure planes, which improves the toughness and strength of cementitious composites [8]. However, the dispersion can be of significant concern in the incorporation of nanomaterials in cement-based materials. Literature has reported a threshold for using nanomaterials that limits the application of these functional materials. Many dispersing techniques have been proposed to reduce the negative impact of the agglomeration of nanomaterials in the matrix. They include physical dispersion methods (e.g., high-speed shear and ultrasonic dispersion), chemical dispersion methods for surface modification (e.g., covalent/noncovalent bond modification), and the combination of physical and chemical processes [11]. The following paragraphs introduce the main inorganic nanomaterials and carbon nanofilaments, their characteristics, and their properties in cement-based materials.

2.2.1 Nanosilica

NS is one of the most widely used nanomaterials to enhance cement-based materials' mechanical properties and durability through physical and chemical effects. Generally, there are two production methods of NS: wet and thermal methods [12]. In the wet method, also known as the water route or sol-gel process, organometallics are used to produce silica nanoparticles [13]. The silicates are treated with mineral acids, and the hydrated NS particles are obtained [14]. The thermal method is based on the vaporization of silica at high temperatures (1500–2000°C). Biological methods [15] and precipitation from olivine mineral dissolution [16] are considered alternative synthesis methods for NS production. The NS is used in different applications, such as drug delivery, tissue engineering, carrier for antimicrobial applications, and biosensing [17], other than construction applications. NS is potentially better than silica fume for its higher content of amorphous silica (>99%) and the smaller particle size (1–50 nm) [18], although the price is still an issue. Hou et al. [19] showed that the role of NS in cement-based materials is mainly chemical, and its reactivity is related to the particle size and its production technique. The formation rate of rigid C—S—H with longer chains increases as silica particles decrease in size [20]. However, the content of NS in the concrete mixture should be controlled more rigorously because of the negative impact of the excessive amount of NS particles (more than 1%) on concrete properties [21].

2.2.2 Nanotitania

NT is produced using different processes in liquid or solid-state, such as solvothermal [22], hydrothermal [23], sol-gel [24], and mechano-chemical [25]. Further information on the synthesis and properties of NT can be found in [26]. NT is used in applications such as photocatalysis, dye-sensitized solar cells, gas sensor, nanomedicine, skincare products, wastewater treatment, and antimicrobial applications [17]. Many researchers have investigated the effects of NT on mechanical and physical properties and the durability of cement paste, mortar, and concrete [27]. The findings indicate that the enhancing effects of NT are mainly due to the chemical and physical properties of the particles. Titania nanoparticles directly affect the hydration rate and time due to a faster dilution of cement compounds and better distribution of hydration products through the heterogeneous nucleation effect, as discussed earlier. Lee and Kurtis [28] found that the additional nucleation sites for hydration products were provided in the cement matrix due to the high surface area of NT particles, which accelerated hydration reactions. They observed the quicker formation of hydration peaks and higher total heat of hydration in both C_3S clinker and Portland cement pastes. The nanofilling effect and the densification of the microstructure were also reported by Farzadnia et al. [29]. However, increasing the percentage of TiO_2 nanoparticles by more than 2% reduced the enhancing effect of NT due to the agglomeration of its nanoparticles [27].

2.2.3 Nanoalumina

NA has shown promising results in improving the mechanical behavior of cement composites despite its inert crystalline structure. NA is synthesized through different processes, including flame spray pyrolysis [30], reverse microemulsion [31], sol-gel [32], precipitation, and freeze-drying [33]. The alumina nanoparticles are used in different applications other than construction, such as wastewater and soil treatment, antimicrobial applications, ceramic ultrafilters and membranes, gas separation, catalysis and absorption processes, and drug delivery [17]. In cement-based materials, studies showed that NA can densify the matrix and reduce the size of pores >1–0.1 μm [34]. This is mainly due to the nanofilling effect of NA particles that can act as ultra-fine aggregate and by filling the pores, which can ultimately affect the modulus of elasticity. Li et al. [9] showed that adding 7% NA increased the elastic modulus of mortar by 208% after 28 days. Besides, the high reactivity of NA particles can increase the hydration and dissolution of C_3S and C_2S phases, resulting in further refinement of the cement matrix [34]. NA particles can improve the engineering properties of cement-based materials, such as compressive strength (by 64%) [35], tensile strength (by 67%) [36], and flexural strength (by 70%) [35]. Farzadnia et al. [37] reported that NA increased residual compressive strength of high-strength concrete after exposure to elevated temperatures due to the formation of Mullite at 800°C. Mullite is a stable phase of aluminosilicate formed at elevated temperatures of about 1150–1350°C [38] due to the reaction between Al_2O_3 and SiO_2. However, incorporating a higher NA dosage (more than 2%) is reported to decrease its performance due to difficulty in dispersion [27].

2.2.4 Carbon-based nanofilaments

The effects of carbon-based nanofilaments, such as carbon nanofibers (CNFs), carbon nanotubes (CNTs), and graphene oxide (GO), in cement matrix are highlighted in various studies due to their enhancing effect on engineering properties and durability of cement-based materials [39,40]. CNFs and multiwalled carbon nanotubes (MWCNTs) are nested arrays of graphene. In contrast, single-walled carbon nanotubes (SWCNTs) are composed of rolled single graphene sheets in the hollow cylindrical form [41]. GO is a derivative of graphene made of an extra hydroxyl (OH) group attached to the same hexagonal carbon network. It is a single-layered nanomaterial with oxygenated graphene sheets that are linked to OH and (—O—) on the base plane and to the carboxyl (COOH) and carbonyl (C=O) groups on the sheet edge [42]. Owing to the hydrophilic nature of the oxygen-functional groups of the GO, the dispersibility of particles is less problematic than those of CNF and CNTs [43,44]. The two-dimensional sheets of GO also offer a higher surface area for nucleation of the hydration products [45,46]. Furthermore, the rough surface structure of sheets leads to better interlocking between GO and hydration products. This improves the bond strength and load transfer efficiency between the cement matrix and GO [47]. For CNFs and CNTs, their superior performance in cement-based materials is originated from their high Young's modulus (400 GPa and 1 TPa for CNF and CNT, respectively) and tensile strength (7 GPa and 60 GP for CNF and CNT, respectively) [48,49]. CNFs/CNTs have a very high aspect ratio, so they bridge the microcracks within the cement matrix and inhibit the propagation of cracks [50]. CNT/CNFs can bridge cracks and voids to form a reinforcing mechanism and obstruct cracking in cement matrix due to their high aspect ratio and nanoscale diameter [51]. Despite the superior performance of carbon-based nanofilaments, there is an increased risk of agglomeration owing to their high specific surface energy (Van der Waals forces) [52]. Hence, their efficiency depends on their state of dispersion [11].

2.3 Surface modification: Methods and techniques

2.3.1 Surface modification of inorganic nanoparticles

Surface modification of inorganic nanoparticles has attracted significant attention because it enhances their efficiency in the nanocomposite. Surface modification improves the dispersibility of nanoparticles in suspensions by providing a strong repulsion between nanoparticles. This can lead to improvements in mechanical properties, rheological and functional properties of nanocomposites. The following sections briefly discuss some of the available approaches for surface modification of inorganic nanoparticles in various applications. In Section 2.4, the specific procedures used by researchers in cement-based materials will be reviewed.

2.3.1.1 Chemical treatments using silane coupling agents

Chemical treatment of inorganic nanoparticles involves chemical reactions to improve the dispersibility of nanoparticles in various suspensions. One good example of chemical treatment is the absorption of silane coupling agents onto the surface of inorganic nanoparticles. Surface modification using silane coupling agents was first introduced by Plueddemann et al. [53]. It was later used to enhance the compatibility between the particle surfaces and suspension and enhance composite materials' properties [54,55]. This method employs the hydrophilic end of silanes such as (octyl) and (n-propyl) tri-ethoxy silanes that can interact with OH groups on the nanoparticle's surface. The modified nanoparticles wrapped with silanes show better dispersion in the suspension. This technique has been reported to be effective for NS [56], nano-Al_2O_3 [57], and NT [58]. Further information about the different silane coupling agents and the underlying mechanisms is available in [17].

2.3.1.2 Grafting synthetic polymers

Another surface modification method for inorganic nanomaterials is grafting synthetic polymers onto the surface of nanoparticles. This method can enhance the chemical functionality and change the surface topology of the particle surfaces. This method uses polymers (monomers) with a very low molecular weight that can react with the active sites on the nanoparticle surface. The interlayers inside the nanoparticle agglomerates can then be filled partially with grafted macromolecular chains. This can cause a repulsive force and separate the nanoparticles. Additionally, grafted polymers onto the surfaces of the nanoparticles make them hydrophobic, which can add to the dispersibility of nanoparticles. This method is the most used approach in the modification of nanoparticles for use in cement-based materials. Two approaches are used in this method: (1) grafting "to" and (2) grafting "from." In the first method, the "grafting to" is done by using functionalized polymers to react with the surface of nanoparticles. In the second method, the polymer chains are grown from a self-assembled monolayer placed onto the nanoparticle surface as an initiating group [59].

Grafting methods have been successfully applied to modify the surface of NS [60–62], NA [63], and NT [64,65]. Different polymers such as polystyrene and polyacrylamide (PAAM) [66], styrene and methyl methacrylate (MMA) [67], polyaniline [64,68], or biocompatible polymer [69] were used. Huang et al. [62] modified the surface of NS with a cationic surfactant, poly (diallyl dimethyl ammonium chloride) (PDDA), for use in cement-based materials. The strong negative charges on the surface of NS were used to graft PDDA. The Si—OH groups on NS particles were converted to $Si—O^-$ by deprotonation under the alkaline condition to produce negatively charged surfaces (at the pH of 10). Then, they used a positively charged PDDA to wrap the NS particles. This was found to be an effective way to reduce the agglomeration of silica nanoparticles and enhance the dispersibility in the suspension. However, such surface modification may affect the reactivity of NS, especially in cement-based materials, as discussed later. Such techniques can be complicated, especially for industrialized production that is needed in concrete construction. Wang et al. [63] used a simple method to modify the surface of NA using a fatty acid as a modifier to enable this process at room temperature. This method can allow the industrialization of modified nanoparticles because of its simplicity. In this method, a measured quantity of modifiers is dissolved and mixed in chloroform at room temperature with a certain amount of NA. The mixture is later filtered, dried, and ground. Finally, the chloroform is evaporated to leave the grafted nanoparticles. The authors suggested that this method is appliable to different modifiers such as stearic acid, hexadecanoic acid, and lauric acid to surface-modify other nanoparticles with a hydrophilic surface.

2.3.1.3 In situ surface modification

In situ surface modification approaches are carried out during the nanoparticle synthesis phase. The ligand exchange technique is one method to modify the surface of nanoparticles with organic solvents at high temperatures [70]. In this process, a ligand layer of a few nanometer-thickness is used to envelop the nanoparticles. This will prevent the transport of free electrons between adjacent nanoparticles and significantly hinder the charge transfer. The thickness of the ligand layers should be ideal enough to limit/allow the transfer of electrons while maintaining the dispersibility of the particles. More suitable ligands can be applied to improve dispersibility. In this method, nanoparticles are grown in the ligand solution by heating the solution to a high temperature (typically 150–320°C). This will transform nanoparticles into chemically active atomic or molecular species and condense to form nanoparticles and capped by ligands [71].

2.3.1.4 Adsorption of polymeric dispersants

Adsorption of polymeric dispersants is one of the most feasible surface modification methods to improve the dispersibility of nanoparticles in suspensions. This method has been used widely to disperse nanomaterials for use in cement-based materials. In this method, highly polar organic solvents such as polycarboxylate (PCE)-based superplasticizers are utilized to disperse the hydrophilic nanoparticles. They can generate steric repulsive forces and increase the surface charge, resulting in improved dispersibility of the nanoparticles. Several studies [27,72] have used this method to incorporate carbon nanofilaments, NS, NT, and NA into cement-based matrices.

2.3.2 Surface modification of carbon-based nanofilaments

CNFs, carbon nanotubes, and graphene dioxide belong to the family of carbon-based nanofilaments, as discussed in Section 2.2. The filaments have good thermal and electrical conductivity and a low linear coefficient of thermal expansion. They can enhance strength, stiffness, and fatigue characteristics and are well suited for applications with exposure to severe conditions such as high temperatures or chemical attacks. However, plain carbon filaments may produce composites with a low bond strength due to weak bonding between the filament and matrix as the surfaces are chemically inert [73]. Surface modification has been widely used to overcome existing problems and make them ideal for smart cementitious composites with self-sensitive ability. This section also includes modifying micro-sized carbon fibers (CFs) due to their extensive use in cement-based materials.

2.3.2.1 Surface modification of carbon fibers

The surface modification of CFs can be implemented using chemical reactions for providing reactive functional groups. This method enables the fibers to generate a good chemical bonding with the matrix. Also, it may include the methods that improve the adhesion by physical entanglement of the fibers in the matrix, such as enhancing the roughness of the surfaces. Both methods can improve the performance of fibers in the matrix by increasing the wettability of the fiber surface, removing the weak boundary layer or contaminants off the surface, physical entanglement of fibers in the matrix, and creating surface porosity consequent increased interlocking effect in the matrix. Applying a thin layer of a coupling agent that can chemically bond to fiber and the matrix is another approach in surface modification of fibers. The following paragraphs introduce some common approaches that are used in the surface modification of CFs. Detailed mechanisms and performance of surface modification of CFs can be found in [74–76].

CFs have strong Van der Waals forces that trigger fibers' agglomeration and result in improper dispersion. Similar to inorganic nanoparticles, grafting polymers on the CNF surface can improve their dispersibility in the matrix. Different polymers were used to graft fibers, such as phenoxybenzoic acid [77] and polyimide chains [78]. In this process, the fibers are functionalized initially and then grafted by polymeric chains. Such treatment can reduce their conductivity compared to nonmodified fibers because of the restricted interaction between fibers induced by the surface grafted polymer chains. Plasma treatment is another method that uses an electronically conducting medium, known as plasma. In this method, nitrogen and oxygen-containing functional groups are introduced on the surfaces of CFs [79,80] to modify their chemical and physical structures. Such modification can increase the fiber-matrix bond strength by producing free radicals, ions, and meta-stable species. Using plasma will remove the surface contaminants, increase fiber surface roughness, add wettability, and deposit functional groups to add to the chemical bonding in some matrices depending on their chemical structure [81]. The oxidation method is another chemical method that includes a set of chemical reactions for the surface modification of CFs. In this method, the carbon filaments are modified by producing acidic functional groups on their surface.

Oxidation is achieved using gases (e.g., air, oxygen, and ozone) or chemical liquids (e.g., nitric acid, hydrochloric acid) [82,83]. The extent of modification and its effectiveness can vary based on the fiber characteristics, molecular structure, and concentration of the oxidative medium, time, and temperature [84,85]. Fig. 2.1 shows the surface-modified CF after 180 min of treatment with HNO_3 compared to the untreated CF. The gamma treatment is used widely to surface-modify CFs to improve their interaction with the composite's matrix. This is an advanced treatment that

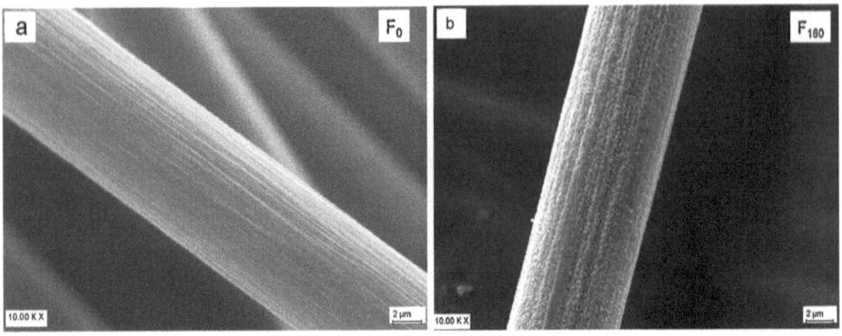

Fig. 2.1 SEM micrograph of untreated and HNO_3 treated CF; (A) untreated (B) 180 min treated CFs [84].

uses high-energy gamma-irradiation or laser irradiation to roughen the surface of filaments. It can also lead to the addition of chemical groups such as carbonyls. Such treatment will enhance the performance of the fiber-reinforced composites by improving the interlocking performance of the fibers [86,87].

2.3.2.2 Surface modification of carbon nanotubes

The modifications of CNTs can be done with two different approaches: (1) involvement of CNTs' structure in a chemical reaction with reagents and introducing chemical groups on the surface through covalent reactions; and (2) adsorption of various molecules onto the CNTs' surfaces using non-covalent interaction. Such techniques can facilitate the well dispersion of CNTs into the suspension and improve the interaction of the nanofibers with the matrix to employ the optimized use of the unique properties of the CNTs.

In covalent modification methods, polymers are attached to the CNT's surfaces using strong covalent chemical bonds. Similar to the modification of inorganic nanomaterials, two approaches are used to graft polymer chains onto the surfaces, "grafting to" and "grafting from" [88,89]. The "grafting to" method involves attaching polymers using the provided functional groups on surfaces of pretreated CNTs [90]. Pretreatment of CNT is mostly carried out using amidation or esterification reactions with the carboxylic acid groups or by derivation of acid chloride groups on CNT's surface. In this approach, polymer chains modified with amino or OH moieties are used to surface-modify CNTs [91]. This method is reliable and straightforward; however, it requires the reactive groups to attach polymers and generates relatively low grafting density. The "grafting from" approach involves in situ polymerization of monomers or attachment of initiators or reactive groups onto the CNTs surface. This approach uses high molecular weight polymers to produce high grafting density. It can utilize the grafting of polymerization and polycondensation of hyperbranched polymers on the pretreated CNTs to improve the dispersibility of CNTs [88,89].

Noncovalent CNT surface modification involves the physical adsorption of branched polymers to the surface of the CNTs. In this method, CNTs are wrapped with conjugated branched polymers or functionalized branched polymers (using π–π-stacking interactions). Modification by π–π-stacking interactions includes functionalizing CNT noncovalently using polycyclic aromatic hydrocarbons (e.g., naphthalene and pyrene) [92]. Such derivates can π-stack to CNTs with high binding energies. Physical interactions can also be used to modify the CNTs by branched polymers or dendrimers. In this approach, physical bonds (entanglement or Van der Waals interactions) are used to attach the polymers to the surface of CNTs. The electrostatic adsorption or hydrogen bonding interactions are also used to surface-modify CNTs by oxidizing nanotubes with acids [93].

2.3.2.3 Surface modification of graphene oxide

Three approaches are used to surface-modify GO: covalent, noncovalent methods, and elemental doping. Similar to CNTs, covalent bond modification of GO involves surface modification of GO with functional groups (covalent bonds) to improve their performance. However, for GO platelets, the oxygen^{-} groups available on their surface provide covalent bonds as their surface contains OH, COOH, and epoxy groups. The main covalent modification of GO includes carbon skeleton modification, hydroxy modification, and COOH modification [94]. In the carbon skeleton modification, the C=C bond in the aromatic ring of GO is mainly used to functionalize the surface [95]. This is a flexible and easy method and is widely used in biosensors. In the hydroxy modification method, a large number of reactive OH groups on the graphene sheet layer are employed to modify the surface. This method involves the hydroxy reaction between GO and amide or isocyanates to produce esters. It facilitates further functionalization of GO using different groups [96]. The COOH modification of GO involves the COOH groups at the edge of GO. COOH modification is a widely used method to modify GO. The COOH group can provide strong bond strength with the modifier due to its high reactivity [97]. This approach includes activating the reaction and forming an ester or amide bond by dehydrating a group containing amino and OH groups.

For noncovalent functionalization, the surface modification of GO mainly involves the interaction between graphene and surfactant functional molecules due to hydrogen bonds and electrostatic forces. This method can maintain the original structure and properties of GO and yet improve their dispersibility. The bonds that facilitate the functional modification of the surface mainly include π–π bond interaction, hydrogen bonding, ionic bonding, and electrostatic interaction modification. The π–π bond interactions refer to aromatic rings in the GO that contain π-bonds. They can generate noncovalent interactions between the surface and the modifier. An excellent example of such surface modification is synergistic interface interactions between GO nanosheets and sulfonated styrene-ethylene/butylene-styrene generated by π–π interaction and hydrogen bonding [98]. In the hydrogen bond interaction, the hydrogen bonding interactions are partly replaced by the interlayer π–π and hydrophobic interactions [99]. In contrast, the carboxylate groups on the graphene surface undergo noncovalent functionalization in the ion interaction methods, using noncovalent ionic interaction with amine-terminated polymers [100].

The electrostatic interaction approach is another technique used to surface-modify GO by creating electrostatic repulsion between the same type of charges on GO's surfaces to improve their dispersion. In this method, the functional groups such as OH groups and epoxy bonds of GO are removed, while the COOH anion remains. The COOH anion provides well dispersion of GO due to charge repulsion [101]. Other than the methods mentioned earlier, element doping modification is applied to surface-modify the GO platelets. This method mainly involves incorporating different elements into graphene by annealing heat treatment, ion bombardment, and arc discharge. This method can reduce graphene defects, maintain the intrinsic structure of graphene, and add new functions to the surface [102]. Using this method can adjust the energy band structure of graphene. However, controlling the doping process quantitatively is complicated.

2.3.2.4 Grafting carbon filaments with nanomaterials

Some efforts have been made to graft nanoparticles on the surface of the carbon nanofilaments through physical and chemical methods. This section reviews the surface modification of CFs to increase fibers' interaction with the matrix and their bridging effect between failed planes. Cui et al. [103] developed a novel grafting method for large-scale applications, using (3-Aminopropyl) tri-ethoxy silane (KH550). In this method, oxidized CF and CNTs were mixed in pure ethanol. Then, the dispersants were added to the solution, stirred, heated in water at 40°C, and sonicated for 30 min. This allowed the dispersant (KH550) to attach onto the surface of oxidized CF and CNTs. The OH groups on the surface of CF and CNTs would react with functional groups in KH550 so that the KH550 could be attached to the surface of oxidized CF and CNTs through chemical bonding [13]. Then, the final products were cooled to room temperature, filtrated and washed thoroughly with deionized water, and dried in an oven at 60°C for 24 h. Fig. 2.2 shows the surface morphologies of CNTs grafted CF. As can be seen, the grafted CF has a tough texture that can increase the interaction with the matrix, especially in cement-based materials. Previously, CF with a smooth surface cannot significantly contribute to the enhanced tensile strength or bending strength [104].

Lu et al. [105] modified the surface of CFs with a thin layer of NS for cement-based materials. The thin layer of NS grafted on the CF could form secondary C—S—H gel by reacting with Ca $(OH)_2$ from the cement matrix. This can improve the engagement of fibers in the matrix through generated interfacial bonds. The sol-gel method was applied to produce

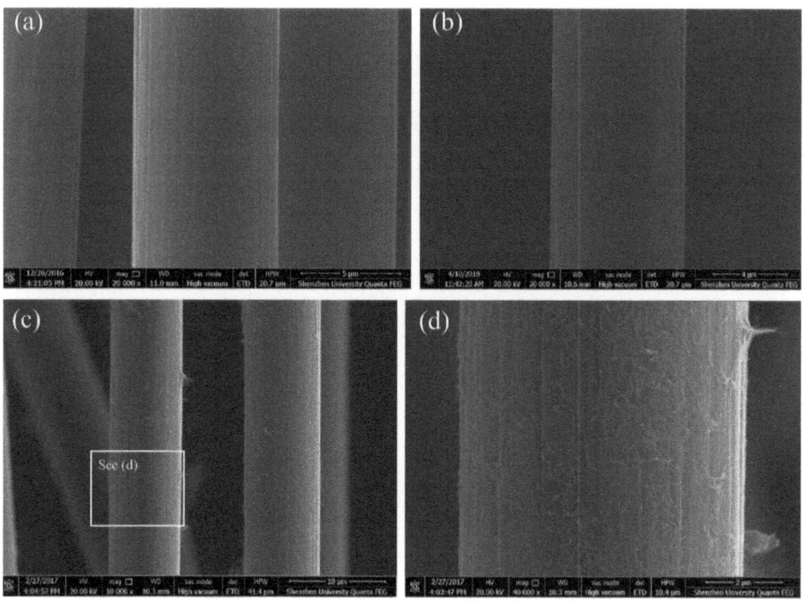

Fig. 2.2 SEM images of CNT grafted CF (A) nonmodified CF, (B) oxidized CF, (C) CF-CNTs, (D) enlarged local surface [103].

NS with high purity at a low temperature. In this approach, the authors used nitric acid, KH550 solution (for preprocessing), ethyl orthosilicate (TEOS), absolute ethyl alcohol (as a solvent), concentrated ammonia (as a catalyst), and cetyltrimethylammonium bromide (CTAB) as a surfactant. The surface modification initially included pretreatment of CF. The fibers were immersed in nitric acid and oxidized for 3 h to enhance their surface activity. Then, the oxidized fiber was maintained in KH550 solution (3%) and washed for 3 h at a constant temperature of 70°C. The final step of the pretreatment was washing the mixture with ethyl alcohol. The final product of the treatment after drying had a higher surface reactivity to host NS particles. The grafting of NS was carried out using pure ethanol as the solvent with the presence of the ionic surfactant (CTAB with a mass fraction of 1%). Then, ammonia was added to provide an alkaline environment. The solution was sonicated for 30 min. Meanwhile, TEOS was added to the mixture and stirred at a low rate under constant temperature (45°C) for 12 h. Such a process led to the formation of insoluble NS particles (a milky white liquid) were formed that were grafted onto the CF surface. Finally, the fibers were washed with methanol and dried at 70°C. Fig. 2.3 shows the surface topology of CNF grafted with NS.

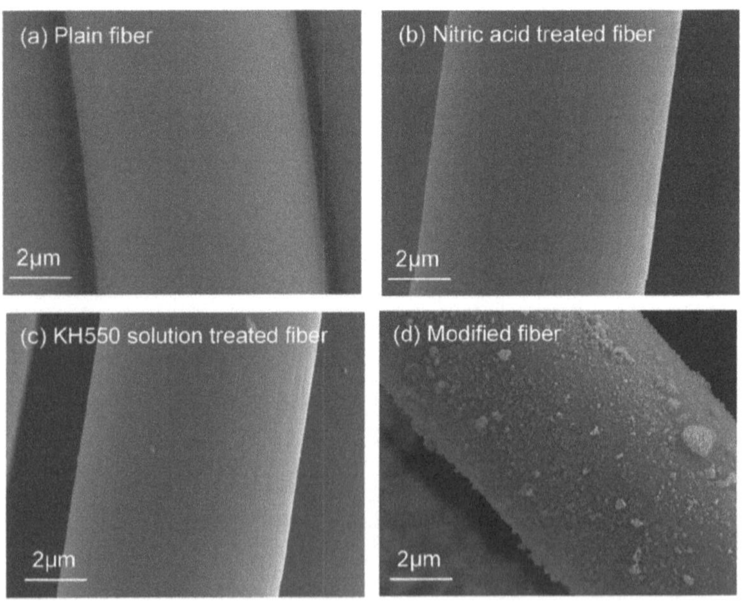

Fig. 2.3 Surface topology of CF at different stages of surface modification [105].

2.3.3 Nanomodification of cement and supplementary cementitious materials

In Sections 2.3.1 and 2.3.2, the surface modification of inorganic nanomaterials and carbon nanofilaments was discussed. This section focuses on the surface modification of cement particles and supplementary cementitious materials (SCMs) using nanomaterials. In such approaches, cement particles, SCM particles, or microfibers are used as carriers for nanomaterials. The ultimate purpose of the surface modification of the cement, SCM particles, and microfibers is to increase the dispersibility of nanoparticles in the cement matrix and to increase the entanglement/interaction of the fibers.

2.3.3.1 Chemical vapor deposition method

The chemical vapor deposition (CVD) method is used to grow CNF/CNT on cement and SCM particles. The CVD method consists of passing gaseous hydrocarbon through the silica-based substrate with the presence of a catalyst exposed to elevated temperatures. The gaseous hydrocarbon will be decomposed on the substrate to produce CNTs/CNFs [106]. This method is also called the hydrocarbon gas pyrolysis method. The effectiveness of this method is based on the substrate type, carbon source, temperature, and catalyst type. It can lead to the formation of different carbon filaments (CNT or CNF) with varying productivity [107].

For the first time, Nasibulin et al. [108] used the CVD method to grow CNTs/CNFs on the surface of cement particles in the carbon source gas environment. No catalyst was used as the substrate (sulfate-resistant cement) as it has an abundant amount of silica, Fe_2O_3, MgO, and Al_2O_3. The carbon source (acetylene) was continuously fed through a container (reactor) where cement particles were exposed as the substrate at a temperature of 550–750°C. This process led to the production of CNFs with a diameter of about 30 nm and an average length of 3 μm on the surface of the cement particles. Mudimela et al. [109] used the same technique to implant CNTs/CNFs on the silica fume particles. The results showed that the in situ growth of the nanofibers (30–50 nm) occurred at a lower temperature of 550°C. The increase in the temperature led to the formation of multiwalled CNTs with 5–10 walls. The outer diameter of multiwalled CNTs varied from 10 to 15 nm at 600°C. The temperature increase to 750°C changed the diameter of nanotubes from 12 nm to 20 nm with a maximum length of 15 μm. The growth time was another critical factor in the growth rate of CNT/CNF. The growth time varied from 6 to 30 min for the optimum growth

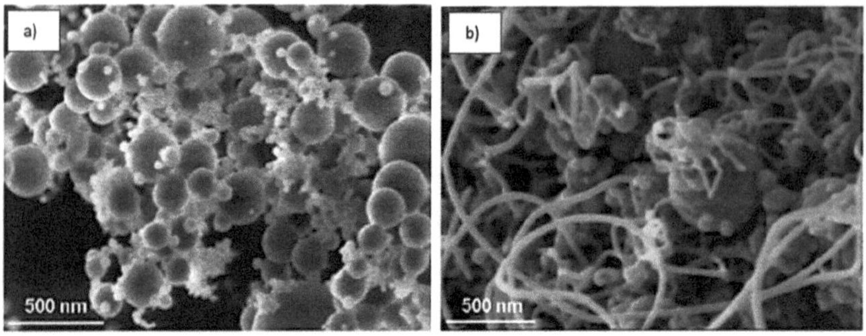

Fig. 2.4 SEM images of (A) silica fume, (B) CNT grown on the surface of silica [109].

condition (1% to 5% of the substrate). Fig. 2.4 shows the silica particles grafted with CNTs.

The in situ growth of CNTs/CNFs on cement and silica fume particles was also carried out by Ludvig et al. [110] using the same technique. However, they applied an external catalyst (Fe_2O_3) using iron ore and industrial by-products (e.g., red mud). The in situ grown CNTs on the cementitious substrates had high purity with multiwalls. The amount of CNT produced using an external catalyst was 10% higher than that of the process without the catalyst [111]. For silica fume particles, the outcome of the process was both CNTs and CNFs with a diameter of 30–60nm [110]. Dunens et al. [112] applied the CVD method for the in situ growth of CNTs/CNFs on fly ash (FA) particles. They also used "Fe" containing external catalyst to synthesize CNTs and CNFs on the surface of particles. For that, the FA was impregnated with iron nitrate to intensify the Fe content. The results showed up to 2.5wt% Fe-loaded FA particles contained 7.65wt% CNTs and CNFs, and the 5wt% Fe-loaded FA products contained 8.25wt% CNTs and CNFs.

2.3.3.2 Microwave irradiating conductive polymer method

The microwave irradiating conductive polymer method is another technique to grow CNT/CNF on cement or SCM particles. Zhang and Liu [113] used the microwave irradiating conductive polymer method to grow CNTs/CNFs on FA particles. This method is faster than CVD and does not need any precautions and extra accessories to protect toxins as no inert gases are required. In this method, microwaves are used to heat a conductive polymer (e.g., polypyrrole) that can absorb irradiation and be used as the heating source to grow carbon nanofilaments. The conductive polymer in solid-state will be mixed with ferrocene powders (iron catalyst) and cyclopentadienyl (carbon source) and heated at above 1100°C using microwave [114]. This method can grow the CNT/CNF on the substrate in 15–30 s [115].

2.3.3.3 Coating SCMs with nanoparticles using a natural binder

For the first time, Farzadnia et al. [116] proposed this approach to coat the surface of rice husk ash (RHA) with industrially manufactured CNFs using chitosan. They used RHA particles as vehicles to disperse CNFs in the cement-based matrix. Chitosan is a natural biopolymer and a chemical derivative obtained by deacetylation of chitin. Chitin is the second-most abundant biopolymer in nature next to cellulose. The coating process of CNFs on RHA included soaking the CNF in an acid solution of HNO_3/H_2SO_4 (70 mL/30 mL) for 48 h to oxidize the CNF surface. This optimum ratio of the acidic solution was chosen by determining the functional groups grafted onto the surface of the CNF [117]. To assure well dispersion of CNF, the CNFs were submerged and stirred in acetone for 24 h before the mixing process of materials. Then, the pretreated CNF and RHA were mixed with the presence of sodium dodecyl sulfate (SDS) surfactant to avoid agglomeration. The mixture of CNF ad RHA was then stirred in a chitosan binder. The chitosan was of a light structure and low molecular weight of 50,000–190,000 Da to provide bonding between CNFs and SiO_2 from RHA. The modified RHA was dried in an oven at 70°C at a relative humidity of <5% for at least 24 h until all the moisture evaporated. In this process, a weight ratio of 10% was used for CNF/RHA. Excessive amounts of CNF showed not to coat the RHA thoroughly and was not cost-effective due to the high price of CNF. Fig. 2.5 shows the SEM image of uncoated and coated RHA with CNF.

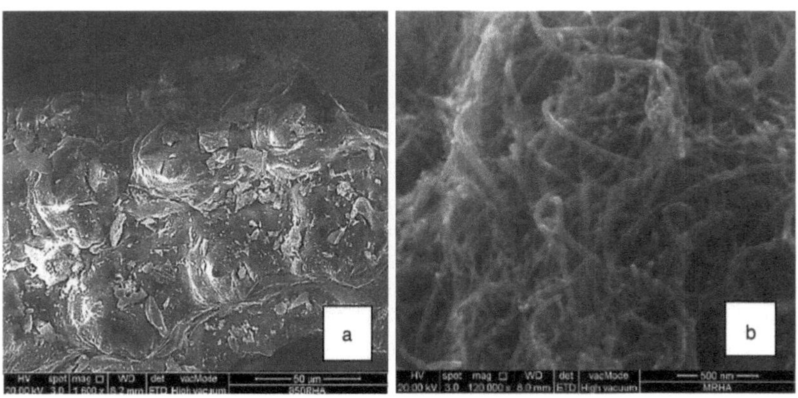

Fig. 2.5 SEM of (A) uncoated and (B) coated RHA with CNF [116].

2.3.3.4 Quick precipitation method

This method is used to synthesize nanoparticles within the porous medium, such as cellulosic fibers. Anggraini et al. [118,119] used this method to plant nanomaterials within the cellulosic structure of coir fibers. This method was used to strengthen the fibers and increase the physical and chemical interactions with the cement matrix. In their work, the precipitation method was used for loading coir fibers on the surface and in the fiber pores with $Fe(OH)_3$, $Al(OH)_3$, $Mg(OH)_2$, and $Ca(OH)_2$ nanoparticles. The method included the quick precipitation of nanoparticles within the pores of cellulosic fiber by immersing the fibers in oversaturated solutions at the ambient temperature/pressure. For example, for precipitation of $Fe(OH)_3$, the coir fibers were soaked in an aqueous solution of 0.5 M $FeCl_3$for 24 h to fill the cellular pores with the $FeCl_3$ solution uniformly. Next, the soaked coir fibers were separated from the $FeCl_3$ aqueous solution and were kept in another beaker. The fibers were kept in an aqueous solution of sodium hydroxide for 24 h. During this treatment, nanoparticles quickly precipitated as nanoparticles on the fiber surfaces and in the cellulosic pores of fibers. The coir fibers were later washed with distilled water to eliminate the residue of the reaction (NaCl and NaOH) from fibers. Then, fibers were dried at ambient room temperature. Fig. 2.6 shows the cellulosic pores before and after precipitations. Results showed that this approach increased the tensile strength of coir fibers by 63%. Bordoloi et al. [120] used the same method to impregnate Fe and Al nanoparticles into other natural fiber structures. The fibers included water hyacinth, jute, and coir. Their microstructural analysis showed successful precipitation of nanoparticles, which led to an increase in the tensile strength of the natural fibers.

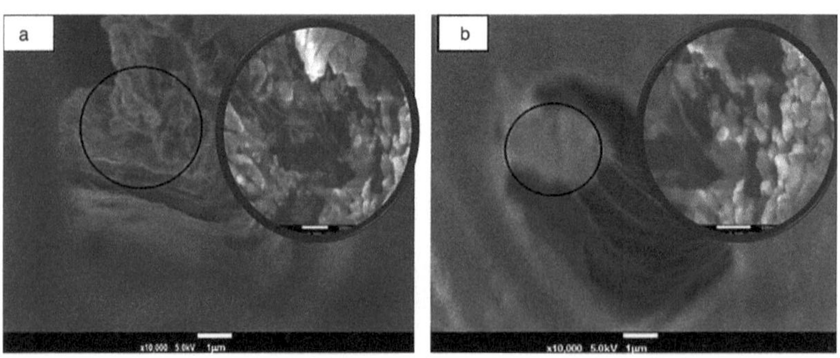

Fig. 2.6 SEM images of (A) unmodified and (B) nanomodified coir fiber microstructure [118].

2.4 Modified nanomaterials in cement-based materials

Nanomodification of cement-based materials is mainly applied to enhance the properties of such composites at the micro- or nano-level. As mentioned in Section 2.2, incorporating inorganic nanomaterials such as NS, NA, and NT can enhance the hydration kinetics, densify the microstructure and shift the pore structure to finer pores. They can collectively improve the engineering properties and durability of cement-based materials. At the same time, such cementitious composites are brittle and weak at a macroscopic level under load stresses (e.g., tensile and shear) or those applied by changes in the temperature and humidity. So, nanoscale fibers such as CNFs [121] and carbon nanotubes [122] are used to overcome this weakness. However, inorganic nanomaterials and carbon filaments are easy to agglomerate due to their high specific surface energy. Therefore, effective surface modifications are needed to disperse them into cement-based materials to implement their enhancing effects fully. Recently, it is realized that surface modification of nanoparticles improves the dispersibility, but it increases the interaction of the particles with the matrix. Some particles have low interaction with the cement matrix due to their surfaces' physical and chemical characteristics. Although many efforts have been carried out to modify nanoparticles, as stated in Section 2.3, studies on the applications of surface-modified nanoparticles in cement-based materials are limited, and some are reviewed in Section 2.4.

2.4.1 Performance of cement-based materials with modified inorganic nanomaterials

Of all surface-modified inorganic nanomaterials, NS has attracted more attention in cement-based materials due to chemical and physical interactions between its particles and the cementitious matrix. Collodetti et al. [123] investigated the surface modification of NS by grafting two types of methoxy-silanes (amine and glycidoxy). The authors investigated the changes in the hydration kinetics and microstructure of cement paste with the modified NS. Fig. 2.7 shows the effect of surface modification of NS on the heat evolution of mixtures. The results showed that the siloxane functional group interfered with the early age reactions of cement pastes and delayed C—S—H formation in the NS/Ca $(OH)_2$ solution. The results also showed that the addition of modified NS to the cement pastes increased the induction period between 15 and 30 h. Besides, the modified NS changed the amount of generated hydration products, namely, C—S—H and

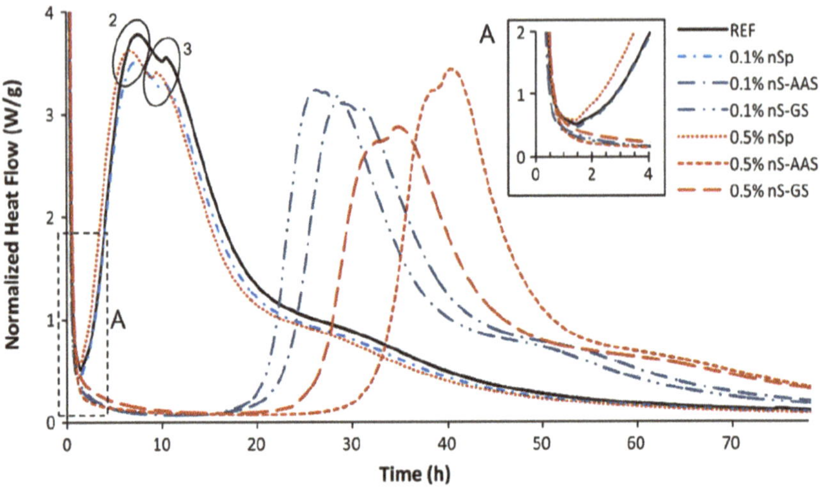

Fig. 2.7 Heat evolution of mixtures with nonmodified NS and surface-modified NS [123].

ettringite. However, the type of modifier had different effects on NS performance in the matrix. A significant effect was observed on pastes with 0.5% amino-based silane-modified NS. However, the modified silica paste and control samples showed similar accumulated heat at 100h of hydration. This may represent a possible application of modified NS to limit the quick render of hydrates at early ages with no significant impact on later ages. The authors related such performance of modified NS to two possible mechanisms: (1) the surface —OH is partly removed or blocked by the grafted polymers [123]. Such OH groups are ionic mediums to react with calcium hydroxide [124]. This behavior of grafted polymers may result in regions on NS with reduced reactivity, (2) a new functional group (from the siloxanes) is added to the NS. This functional group may determine the type of functional coverage and the interaction with cement paste. So, using a proper modifier that can increase the chemical interaction with Ca $(OH)_2$ can be instrumental in the hydration kinetics.

Huang et al. [62] studied the potential application of surface-modified NS in high-performance cement paste made with a water-to-binder (w/b) of 0.35. The authors used PDDA as the modifier. The ultimate purpose of their study was to investigate the effect of surface modification of NS on hydration kinetics. (See Fig. 2.1 in Section 2.3.1.2 for the schematics of surface modification of NS and the modification process.) Results showed that the surface modification of NS with PDDA retarded the hydration and

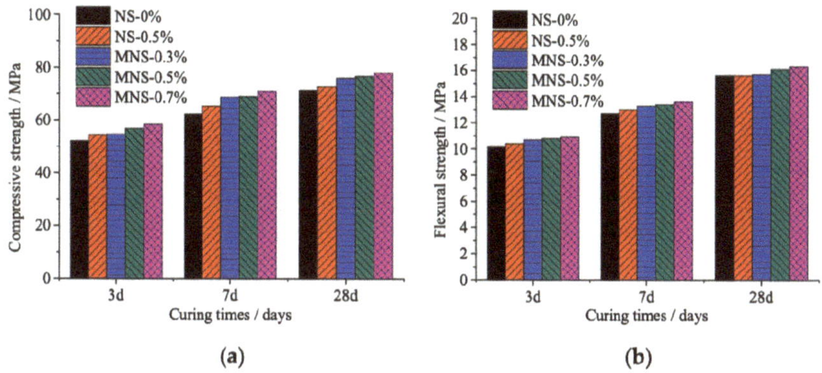

Fig. 2.8 (A) Compressive strength and (B) flexural strength of HPC at different curing ages made with nonmodified and modified NS at different dosages [61].

delayed the exothermic peaks of cement hydration. The addition of higher PDDA modified NS contents further delayed the formation of the main hydration peak and lowered the released heat. However, Rong et al. [61] showed that surface modification of NS using silane coupling agent (KH550) as the modifying agent was beneficial to NS's hydration kinetics. The authors studied the effect of surface-modified NS on the mechanical properties of high-performance cement-based composites. The results showed that the modified NS accelerated the hydration of cement. Also, the pozzolanic filling and nucleation effects of the modified NS led to a denser microstructure of the matrix and enhanced mechanical properties of the cement-based composites. Fig. 2.8 shows the compressive and flexural strengths of samples with modified and nonmodified NS. The optimum modification parameters represented a coupling agent content of 10%, reaction temperature of 65°C, and reaction time of 8 h. Compared with the nonmodified NS, such parameters prevented the condensation of silane alcohol groups between nanoparticles to reduce their reactivity. The modified NS also showed better dispersibility in saturated calcium hydroxide solution. The authors reported that such improvement is due to the coupling agent grafted on the surface of nanoparticles that caused the steric hindrance and prevented their agglomeration [61].

2.4.2 Performance of cement-based materials with modified carbon filaments

As mentioned in Section 2.3, the surface modification of carbon filament can be done by applying chemical reactions. Such reactions provide reactive functional groups on the surface. Also, grafting filaments with nanomaterials can improve the physical and chemical entanglement of the nanofilaments. Xu and Chung [125] studied the effect of surface-modified CFs on the mechanical properties of cement paste. The authors modified CFs using three different surface treatment methods using oxidation by ozone and an aqueous solution of potassium dichromate and sulfuric acid to enhance the oxidation ability and silane grafting. The results showed that the tensile strength and modulus of samples increased by 26% and 14%, respectively, and the air void content decreased when the CF was surface modified. The silane-treated CF worked more efficiently compared to other modification techniques. The effectiveness of treatment decreased in the order: silane-treated fibers, dichromate-treated fibers, and ozone-treated fibers. The positive effects of silane treatment were attributed to the hydrophilic nature of silane. Also, the drying shrinkage of cement paste decreased in the samples with silane-modified CF which can be related to the increased hydrophilicity of fibers after the treatment and the formation of chemical bonds between fibers and cement. The shrinkage of samples at 28 days decreased by 32%.

Such hydrophilic character of the silane-modified CF also improved the bond strength between fibers and cement paste and increased the tensile strength, modulus and ductility. Li et al. [126] investigated the effect of surface-modified CNT in the cement matrix. They modified the surface of CNT using H_2SO_4 and HNO_3 and studied the mechanical properties and microstructure of the mixtures. The results showed that adding modified carbon nanotubes to the cement paste enhanced the flexural and compressive strengths by 25% and 19%, respectively. This was related to the enhancement in the interfacial interactions by a strong covalent force between surface-modified nanotubes and hydrates, such as C—S—H and calcium hydroxide, increasing the high bonding strength. Also, the results showed that the addition of carbon nanotubes shifted the pore structure to finer pores. They reported that using sulfuric acid and nitric acid to generate carboxylic acid groups on the surfaces of carbon nanotubes triggered chemical reactions between the carboxylic acid and the C—S—H or $Ca(OH)_2$. Fig. 2.9 shows the SEM images of the interfacial transition zone (ITZ) of surface-modified CF in the cement matrix.

Cui et al. [103] studied the effect of CF grafted with CNT on the mechanical properties of cement paste. The results showed that the CNF increased the flexural strengths of cement paste at 3d, 7d, and 28d by 49%, 42%, and 46%, respectively, compared to the control samples. Grafting

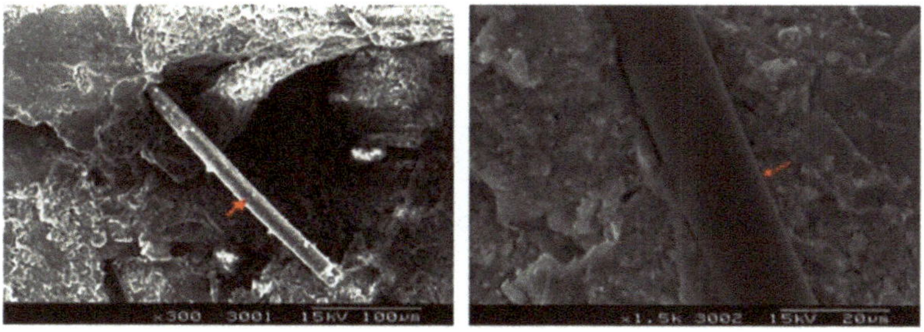

Fig. 2.9 SEM image of CF cement matrix composites (*Arrows*: carbon fibers) [126].

increased flexural strength by 21%, 15%, and 22% compared to samples with nonmodified CF. They related such enhancement to the improved interfacial bond strength between CF and the cement matrix. They also stated that the chemical reaction between the surface functional groups and the cement hydration products increased the flexural strength. Fig. 2.10 shows the improvement in the flexural strength of samples with modified CF. However, the grafted CF had an insignificant effect on the compressive strength of samples.

Lu et al. [105] studied the effect of NS grafted CF on the ITZ properties of fiber and cement matrix in fiber-reinforced cementitious composites. The surface modification process is explained in Section 2.3.2.4. The results show that the NS particles grafted on the surface of CNFs reacted with the cement hydration product, Ca $(OH)_2$, to form a C-S-H gel. Such a reaction increased the bond strength at the ITZ of the fiber and improved the mechanical properties of the mixtures. The pullout test results showed that the bond strength, frictional bond strength, and chemical debonding energy of modified fiber in cement matrix were approximately 3, 2, and 10 times higher than the samples with nonmodified CNF.

Some studies also investigated the effect of nanomodification of cement and SCM particles on the properties of the cementitious matrix. Nasibulin et al. [127] studied the effect of in situ growth of CNTs/CNFs on cement particles on the properties of cement mortars. The properties included compressive strength, flexural strength, and electrical resistance. The mechanical properties showed that the addition CNTs/CNFs modified cement particles and decreased the early age compressive strength of samples due to the lower degree of hydration of the modified cement particles. However, the samples prepared with CNTs/CNFs modified cement had compressive strength two folds higher than that of control samples at later ages. The electrical

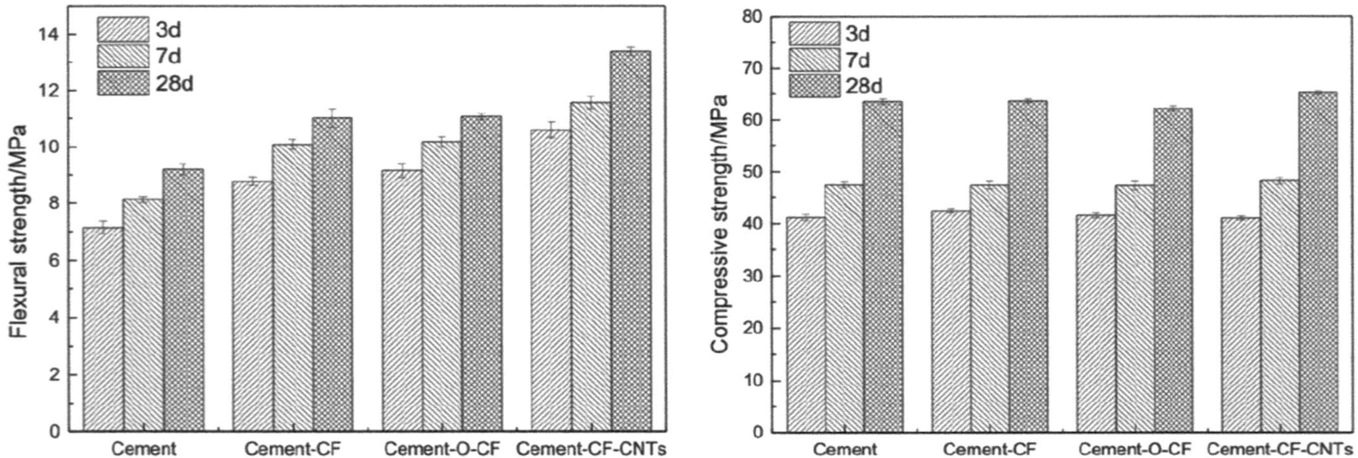

(a) Flexural strength (b) Compressive strength

Fig. 2.10 mechanical properties of cement paste reinforced with surface-modified CF with CNT [103].

conductivity of those samples with modified cement particles showed a 70-time increase. Hlavacek et al. [128] also used CNTs/CNFs modified cement particles in pastes with the w/b of 0.35 and the CNTs/CNFs paste ratio up to 0.038% by weight. The results showed 25% increase in compressive strength and 14% increase in fracture energy with the addition of 3.5% CNTs/CNFs modified cement. The CVD technique was also used in a study by Ludvig et al. [110]. The results also indicated significant improvement in the mechanical properties of samples made with modified cement particles. Results showed a 35% increase in the tensile strength, 14% higher flexural strength, and 88% increase in compressive strength of mortar samples when 0.3% of CNTs/CNFs modified cement particle was used. Farzadnia et al. [116] studied the effect of coated RHA with CNFs on the mechanical properties of cement paste. The detailed coating process is described in Section 2.3.3.3. The results indicated that the paste with coated RHA had a 53% and 187% increase in compressive and flexural strengths, respectively, at 28 days. Also, the inclusion of modified RHA in samples improved the ductility, increased the modulus of elasticity and toughness, and shifted the failure mode from strain softening (quasibrittleness) to pseudo strain hardening. Fig. 2.11 shows the flexural properties of cement paste with CNF modified and nonmodified RHA.

2.5 Conclusions and future perspectives

This chapter reviewed some modification techniques of nanoparticles and carbon filament for different applications and emphasized their use in cement-based materials. Such techniques included chemical approaches, including oxidation treatment using acids or silane coupling agents, grafting polymers onto the surface of nanoparticles, and in situ surface modification during the synthesis of nanoparticles. Also, some physical approaches, such as applying high-energy gamma-irradiation or employing the interaction between hydrogen bonds and electrostatic forces between particles and surfactant functional molecules, were reviewed. Such physical approaches are used to increase the entanglement of fibers in the matrix. At the same time, the surface modification, either chemical or physical, can improve the dispersibility of nanoparticles and carbon filaments and lead to their higher efficiency in the composite. This chapter also studied the surface modification of CFs (microscale) using nanomaterials such as NS and carbon nanotubes. Some techniques were also discussed to nano-modify cement and cementitious binder particles such as silica fume, FA, and rusk husk ash.

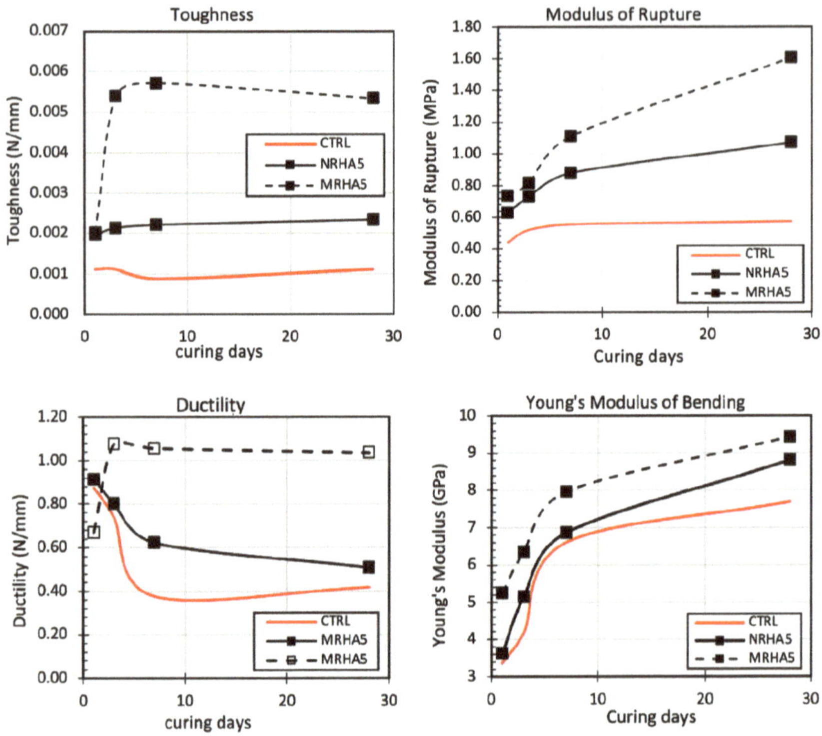

Fig. 2.11 Flexural properties of paste with CNF coated RHA [116].

This chapter also reviewed some surface modification methods used for inorganic nanoparticles and carbon filaments in cement-based materials, and their effect on performance was studied. The results showed that modified nanoparticles can improve the mechanical properties of cementitious composites; however, surface modification can reduce the reactivity of particles. This is more significant in nanoparticles such as NS, which plays an important role in chemical reactivity. In general, surface modification parameters, such as thickness, electronic charges, duration, and temperature of the modification process and the modifier (e.g., polymers), should be controlled. Despite many studies that have been conducted in the area of modification of nanomaterials for different applications, minimal knowledge is available on best-suited methods to be used in cement-based materials.

References

[1] A. Zimmer, M.J. de Andrade, F.A.L. Sánchez, A.S. Takimi, Use of natural and modified natural nanostructured materials, in: Nanostructured Materials for Engineering Applications, Springer, 2011, pp. 157–172.

[2] A.S. Edelstein, R. Cammaratra, Nanomaterials: Synthesis, Properties and Applications, CRC press, 1998.
[3] M. Faraday, X. The Bakerian Lecture.—Experimental relations of gold (and other metals) to light, Philos. Trans. R. Soc. Lond. (1857) 145–181.
[4] G.A. Ozin, A. Arsenault, Nanochemistry: A Chemical Approach to Nanomaterials, Royal Society of Chemistry, 2015.
[5] S. Nazar, J. Yang, B.S. Thomas, I. Azim, S.K.U. Rehman, Rheological properties of cementitious composites with and without nanomaterials: a comprehensive review, J. Clean. Prod. (2020), 122701.
[6] K. Sobolev, I. Flores, L. Torres-Martinez, P. Valdez, E. Zarazua, E. Cuellar, Engineering of SiO_2 nanoparticles for optimal performance in nano cement-based materials, in: Nanotechnology in Construction 3, Springer, 2009, pp. 139–148.
[7] K. Sobolev, S.P. Shah, Nanotechnology in Construction: Proceedings of NICOM5, Springer, 2015.
[8] K. Gopalakrishnan, B. Birgisson, P. Taylor, N.O. Attoh-Okine, Nanotechnology in Civil Infrastructure: A Paradigm Shift, Springer, 2011.
[9] Z. Li, H. Wang, S. He, Y. Lu, M. Wang, Investigations on the preparation and mechanical properties of the nano-alumina reinforced cement composite, Mater. Lett. 60 (2006) 356–359.
[10] Z. Bittnar, P.J. Bartos, J. Nemecek, V. Smilauer, J. Zeman, Nanotechnology in Construction: Proceedings of the NICOM3, Springer Science & Business Media, 2009.
[11] A. Korayem, N. Tourani, M. Zakertabrizi, A. Sabziparvar, W. Duan, A review of dispersion of nanoparticles in cementitious matrices: nanoparticle geometry perspective, Construct. Build Mater. 153 (2017) 346–357.
[12] G. Quercia, H. Brouwers, Application of Nano-Silica (nS) in Concrete Mixtures, 2010, pp. 431–436.
[13] W. Stöber, A. Fink, E. Bohn, Controlled growth of monodisperse silica spheres in the micron size range, J. Colloid Interface Sci. 26 (1968) 62–69.
[14] H. Zou, S. Wu, J. Shen, Polymer/silica nanocomposites: preparation, characterization, properties, and applications, Chem. Rev. 108 (2008) 3893–3957.
[15] M. Estevez, S. Vargas, V. Castano, R. Rodriguez, Silica nanoparticles produced by worms through a bio-digestion process of rice husk, J. Non Cryst. Solids 355 (2009) 844–850.
[16] D.J. Lieftink, The Preparation and Characterization of Silica from Acid Treatment of Olivine, Universiteit Utrecht, 1997.
[17] S. Kango, S. Kalia, A. Celli, J. Njuguna, Y. Habibi, R. Kumar, Surface modification of inorganic nanoparticles for development of organic–inorganic nanocomposites—a review, Prog. Polym. Sci. 38 (2013) 1232–1261.
[18] Y. Qing, Z. Zenan, K. Deyu, C. Rongshen, Influence of nano-SiO_2 addition on properties of hardened cement paste as compared with silica fume, Construct. Build Mater. 21 (2007) 539–545.
[19] P. Hou, J. Qian, X. Cheng, S.P. Shah, Effects of the pozzolanic reactivity of nanoSiO_2 on cement-based materials, Cem. Concr. Compos. 55 (2015) 250–258.
[20] J.S. Belkowitz, The Investigation of Nano Silica in the Cement Hydration Pro, 2009.
[21] A.M. Rashad, A comprehensive overview about the effect of nano-SiO_2 on some properties of traditional cementitious materials and alkali-activated fly ash, Construct. Build Mater. 52 (2014) 437–464.
[22] C.-S. Kim, B.K. Moon, J.-H. Park, S.T. Chung, S.-M. Son, Synthesis of nanocrystalline TiO_2 in toluene by a solvothermal route, J. Cryst. Growth 254 (2003) 405–410.
[23] J.-N. Nian, H. Teng, Hydrothermal synthesis of single-crystalline anatase TiO_2 nanorods with nanotubes as the precursor, J. Phys. Chem. B 110 (2006) 4193–4198.

[24] T. Trung, W.-J. Cho, C.-S. Ha, Preparation of TiO_2 nanoparticles in glycerol-containing solutions, Mater. Lett. 57 (2003) 2746–2750.
[25] M. Kamei, T. Mitsuhashi, Hydrophobic drawings on hydrophilic surfaces of single crystalline titanium dioxide: surface wettability control by mechanochemical treatment, Surf. Sci. 463 (2000) L609–L612.
[26] X. Chen, S.S. Mao, Titanium dioxide nanomaterials: synthesis, properties, modifications, and applications, Chem. Rev. 107 (2007) 2891–2959.
[27] M. Sumesh, U.J. Alengaram, M.Z. Jumaat, K.H. Mo, M.F. Alnahhal, Incorporation of nanomaterials in cement composite and geopolymer based paste and mortar–a review, Construct. Build Mater. 148 (2017) 62–84.
[28] B.Y. Lee, K.E. Kurtis, Influence of TiO_2 nanoparticles on early C3S hydration, J. Am. Ceram. Soc. 93 (2010) 3399–3405.
[29] N. Farzadnia, A.A.A. Ali, R. Demirboga, M.P. Anwar, Characterization of high strength mortars with nano Titania at elevated temperatures, Construct. Build Mater. 43 (2013) 469–479.
[30] A. Tok, F. Boey, X. Zhao, Novel synthesis of Al_2O_3 nano-particles by flame spray pyrolysis, J. Mater. Process. Technol. 178 (2006) 270–273.
[31] M.A. Malik, M.Y. Wani, M.A. Hashim, Microemulsion method: a novel route to synthesize organic and inorganic nanomaterials: 1st nano update, Arab. J. Chem. 5 (2012) 397–417.
[32] M. Farahmandjou, N. Golabiyan, New pore structure of nano-alumina (Al_2O_3) prepared by sol gel method, J. Ceram. Process. Res. 16 (2015) 1–4.
[33] J. Xiao, Y. Wan, H. Deng, J. Li, Y. Liu, Effects of drying method on preparation of nanometer α-Al_2O_3, J. Cent. South Univ. Technol. 14 (2007) 330–335.
[34] I. Campillo, A. Guerrero, J. Dolado, A. Porro, J.A. Ibáñez, S. Goñi, Improvement of initial mechanical strength by nanoalumina in belite cements, Mater. Lett. 61 (2007) 1889–1892.
[35] M. Arefi, M. Javeri, E. Mollaahmadi, To study the effect of adding Al_2O_3 nanoparticles on the mechanical properties and microstructure of cement mortar, Life Sci. J. 8 (2011) 613–617.
[36] A. Nazari, S. Riahi, S. Riahi, S.F. Shamekhi, A. Khademno, Mechanical properties of cement mortar with Al_2O_3 nanoparticles, J. Am. Sci. 6 (2010) 94–97.
[37] N. Farzadnia, A.A.A. Ali, R. Demirboga, Characterization of high strength mortars with nano alumina at elevated temperatures, Cem. Concr. Res. 54 (2013) 43–54.
[38] M.D. Sacks, K. Wang, G.W. Scheiffele, N. Bozkurt, Effect of composition on mullitization behavior of α-alumina/silica microcomposite powders, J. Am. Ceram. Soc. 80 (1997) 663–672.
[39] F. Sanchez, Carbon nanofibre/cement composites: challenges and promises as structural materials, Int. J. Mater. Struct. Integr. 3 (2009) 217–226.
[40] T. Mendes, D. Hotza, W. Repette, Nanoparticles in cement based materials: a review, Rev. Adv. Mater. Sci. 40 (2015) 89–96.
[41] T. Shi, Z. Li, J. Guo, H. Gong, C. Gu, Research progress on CNTs/CNFs-modified cement-based composites – a review, Construct. Build Mater. 202 (2019) 290–307.
[42] D. Chen, H. Feng, J. Li, Graphene oxide: preparation, functionalization, and electrochemical applications, Chem. Rev. 112 (2012) 6027–6053.
[43] Y.S. Yun, H.-J. Jin, Difference of dispersion behavior between graphene oxide and oxidized carbon nanotubes in polar organic solvents, Curr. Appl. Phys. 12 (2012) 637–642.
[44] L. Qiu, X. Yang, X. Gou, W. Yang, Z. Ma, G.G. Wallace, D. Li, Dispersing carbon nanotubes with graphene oxide in water and synergistic effects between graphene derivatives, Chem. A Eur. J. 16 (2010) 10653–10658.
[45] F. Babak, H. Abolfazl, R. Alimorad, G. Parviz, Preparation and mechanical properties of graphene oxide: cement nanocomposites, Sci. World J. 2014 (2014).

[46] K. Gong, Z. Pan, A.H. Korayem, L. Qiu, D. Li, F. Collins, C.M. Wang, W.H. Duan, Reinforcing effects of graphene oxide on Portland cement paste, J. Mater. Civ. Eng. 27 (2015), A4014010.

[47] Y. Xu, J. Zeng, W. Chen, R. Jin, B. Li, Z. Pan, A holistic review of cement composites reinforced with graphene oxide, Construct. Build Mater. 171 (2018) 291–302.

[48] Z. Zhou, C. Lai, L. Zhang, Y. Qian, H. Hou, D.H. Reneker, H. Fong, Development of carbon nanofibers from aligned electrospun polyacrylonitrile nanofiber bundles and characterization of their microstructural, electrical, and mechanical properties, Polymer 50 (2009) 2999–3006.

[49] M.-F. Yu, O. Lourie, M.J. Dyer, K. Moloni, T.F. Kelly, R.S. Ruoff, Strength and breaking mechanism of multi-walled carbon nanotubes under tensile load, Science 287 (2000) 637–640.

[50] J.-P. Salvetat, J.-M. Bonard, N. Thomson, A. Kulik, L. Forro, W. Benoit, L. Zuppiroli, Mechanical properties of carbon nanotubes, Appl. Phys. A 69 (1999) 255–260.

[51] J. Makar, J. Margeson, J. Luh, Carbon Nanotube/Cement Composites-Early Results and Potential Applications, Vancouver Canada, 2005, pp. 1–10.

[52] B.M. Tyson, R.K.A. Al-Rub, A. Yazdanbakhsh, Z. Grasley, Carbon nanotubes and carbon nanofibers for enhancing the mechanical properties of nanocomposite cementitious materials, J. Mater. Civ. Eng. 23 (2011) 1028–1035.

[53] E.P. Plueddemann, H. Clark, L. Nelson, K. Hoffmann, Silane coupling agents for reinforced plastics, Mod. Plast. 39 (1962) 135.

[54] E.P. Plueddemann, Adhesion through silane coupling agents, J. Adhes. 2 (1970) 184–201.

[55] M.J. Owen, Coupling agents: chemical bonding at interfaces, in: Adhesion Science and Engineering, Elsevier, 2002, pp. 403–431.

[56] Y. Guo, M. Wang, H. Zhang, G. Liu, L. Zhang, X. Qu, The surface modification of nanosilica, preparation of nanosilica/acrylic core-shell composite latex, and its application in toughening PVC matrix, J. Appl. Polym. Sci. 107 (2008) 2671–2680.

[57] L.T. Truong, Å. Larsen, B. Holme, S. Diplas, F.K. Hansen, J. Roots, S. Jørgensen, Dispersibility of silane-functionalized alumina nanoparticles in syndiotactic polypropylene, Surf. Interface Anal. 42 (2010) 1046–1049.

[58] S. Mallakpour, A. Barati, Efficient preparation of hybrid nanocomposite coatings based on poly (vinyl alcohol) and silane coupling agent modified TiO_2 nanoparticles, Prog. Org. Coat. 71 (2011) 391–398.

[59] O. Prucker, J. Rühe, Synthesis of poly (styrene) monolayers attached to high surface area silica gels through self-assembled monolayers of azo initiators, Macromolecules 31 (1998) 592–601.

[60] Y. Tai, J. Qian, Y. Zhang, J. Huang, Study of surface modification of nano-SiO_2 with macromolecular coupling agent (LMPB-g-MAH), Chem. Eng. J. 141 (2008) 354–361.

[61] Z. Rong, M. Zhao, Y. Wang, Effects of modified nano-SiO_2 particles on properties of high-performance cement-based composites, Materials 13 (2020) 646.

[62] C. Huang, Y. Wang, J. Zhao, D. Wang, Potential effect of surface modified Nano-SiO_2 with PDDA on the cement paste early hydration, ChemistrySelect 5 (2020) 3159–3163.

[63] Y. Wang, W. Eli, L. Zhang, H. Gao, Y. Liu, P. Li, A new method for surface modification of nano-$CaCO_3$ and nano-Al_2O_3 at room temperature, Adv. Powder Technol. 21 (2010) 203–205.

[64] X. Li, W. Chen, C. Bian, J. He, N. Xu, G. Xue, Surface modification of TiO_2 nanoparticles by polyaniline, Appl. Surf. Sci. 217 (2003) 16–22.

[65] M. Sabzi, S. Mirabedini, J. Zohuriaan-Mehr, M. Atai, Surface modification of TiO_2 -nano-particles with silane coupling agent and investigation of its effect on the properties of polyurethane composite coating, Prog. Org. Coat. 65 (2009) 222–228.

[66] M.Z. Rong, Q.L. Ji, M.Q. Zhang, K. Friedrich, Graft polymerization of vinyl monomers onto nanosized alumina particles, Eur. Polym. J. 38 (2002) 1573–1582.
[67] A. Sidorenko, S. Minko, G. Gafijchuk, S. Voronov, Radical polymerization initiated from a solid substrate. 3. Grafting from the surface of an ultrafine powder, Macromolecules 32 (1999) 4539–4543.
[68] X. Li, G. Wang, X. Li, Surface modification of nano-SiO_2 particles using polyaniline, Surf. Coat. Technol. 197 (2005) 56–60.
[69] R. Yokoyama, S. Suzuki, K. Shirai, T. Yamauchi, N. Tsubokawa, M. Tsuchimochi, Preparation and properties of biocompatible polymer-grafted silica nanoparticle, Eur. Polym. J. 42 (2006) 3221–3229.
[70] C.B. Murray, D.J. Norris, M.G. Bawendi, Synthesis and characterization of nearly monodisperse CdE (E= sulfur, selenium, tellurium) semiconductor nanocrystallites, J. Am. Chem. Soc. 115 (1993) 8706–8715.
[71] W.U. Huynh, J.J. Dittmer, W.C. Libby, G.L. Whiting, A.P. Alivisatos, Controlling the morphology of nanocrystal–polymer composites for solar cells, Adv. Funct. Mater. 13 (2003) 73–79.
[72] H. Shao, B. Chen, B. Li, S. Tang, Z. Li, Influence of dispersants on the properties of CNTs reinforced cement-based materials, Construct. Build Mater. 131 (2017) 186–194.
[73] J. Moosburger-Will, J. Jäger, J. Strauch, M. Bauer, S. Strobl, F.F. Linscheid, S. Horn, Interphase formation and fiber matrix adhesion in carbon fiber reinforced epoxy resin: influence of carbon fiber surface chemistry, Compos. Interfaces 24 (2017) 691–710.
[74] S. Tiwari, J. Bijwe, Surface treatment of carbon fibers – a review, Procedia Technology. 14 (2014) 505–512.
[75] N. Raphael, K. Namratha, B. Chandrashekar, K.K. Sadasivuni, D. Ponnamma, A. Smitha, S. Krishnaveni, C. Cheng, K. Byrappa, Surface modification and grafting of carbon fibers: a route to better interface, Prog. Cryst. Growth Charact. Mater. 64 (2018) 75–101.
[76] N. Roy, R. Sengupta, A.K. Bhowmick, Modifications of carbon for polymer composites and nanocomposites, Prog. Polym. Sci. 37 (2012) 781–819.
[77] J.-B. Baek, C.B. Lyons, L.-S. Tan, Grafting of vapor-grown carbon nanofibers via in-situ polycondensation of 3-phenoxybenzoic acid in poly (phosphoric acid), Macromolecules 37 (2004) 8278–8285.
[78] M.J. Arlen, D. Wang, J.D. Jacobs, R. Justice, A. Trionfi, J.W. Hsu, D. Schaffer, L.-S. Tan, R.A. Vaia, Thermal-electrical character of in situ synthesized polyimide-grafted carbon nanofiber composites, Macromolecules 41 (2008) 8053–8062.
[79] M. Heintze, V. Brüser, W. Brandl, G. Marginean, H. Bubert, S. Haiber, Surface functionalisation of carbon nano-fibres in fluidised bed plasma, Surf. Coat. Technol. 174 (2003) 831–834.
[80] V. Brüser, M. Heintze, W. Brandl, G. Marginean, H. Bubert, Surface modification of carbon nanofibres in low temperature plasmas, Diamond Relat. Mater. 13 (2004) 1177–1181.
[81] R. Li, L. Ye, Y.-W. Mai, Application of plasma technologies in fibre-reinforced polymer composites: a review of recent developments, Compos. A: Appl. Sci. Manuf. 28 (1997) 73–86.
[82] L. Lavagna, S. Musso, G. Ferro, M. Pavese, Cement-based composites containing functionalized carbon fibers, Cem. Concr. Compos. 88 (2018) 165–171.
[83] X. Chen, K. Yoon, C. Burger, I. Sics, D. Fang, B.S. Hsiao, B. Chu, In-situ X-ray scattering studies of a unique toughening mechanism in surface-modified carbon nanofiber/UHMWPE nanocomposite films, Macromolecules 38 (2005) 3883–3893.
[84] S. Tiwari, J. Bijwe, S. Panier, Tribological studies on polyetherimide composites based on carbon fabric with optimized oxidation treatment, Wear 271 (2011) 2252–2260.

[85] D. Sebastián, I. Suelves, R. Moliner, M. Lázaro, The effect of the functionalization of carbon nanofibers on their electronic conductivity, Carbon 48 (2010) 4421–4431.
[86] J. Li, Y. Huang, Z. Xu, Z. Wang, High-energy radiation technique treat on the surface of carbon fiber, Mater. Chem. Phys. 94 (2005) 315–321.
[87] J. George, A.K. Bhowmick, Fabrication and properties of ethylene vinyl acetate-carbon nanofiber nanocomposites, Nanoscale Res. Lett. 3 (2008) 508–515.
[88] Z. Spitalsky, D. Tasis, K. Papagelis, C. Galiotis, Carbon nanotube–polymer composites: chemistry, processing, mechanical and electrical properties, Prog. Polym. Sci. 35 (2010) 357–401.
[89] P. Liu, Modifications of carbon nanotubes with polymers, Eur. Polym. J. 41 (2005) 2693–2703.
[90] Y. You, C. Hong, C. Pan, Preparation of smart polymer/carbon nanotube conjugates via stimuli-responsive linkages, Adv. Funct. Mater. 17 (2007) 2470–2477.
[91] G. Jiang, L. Wang, C. Chen, X. Dong, T. Chen, H. Yu, Study on attachment of highly branched molecules onto multi-walled carbon nanotubes, Mater. Lett. 59 (2005) 2085–2089.
[92] R.J. Chen, Y. Zhang, D. Wang, H. Dai, Noncovalent sidewall functionalization of single-walled carbon nanotubes for protein immobilization, J. Am. Chem. Soc. 123 (2001) 3838–3839.
[93] H. Deka, N. Karak, R.D. Kalita, A.K. Buragohain, Biocompatible hyperbranched polyurethane/multi-walled carbon nanotube composites as shape memory materials, Carbon 48 (2010) 2013–2022.
[94] W. Yu, L. Sisi, Y. Haiyan, L. Jie, Progress in the functional modification of graphene/graphene oxide: a review, RSC Adv. 10 (26) (2020) 15328–15345.
[95] A. Sinitskii, A. Dimiev, D.A. Corley, A.A. Fursina, D.V. Kosynkin, J.M. Tour, Kinetics of diazonium functionalization of chemically converted graphene nanoribbons, ACS Nano 4 (2010) 1949–1954.
[96] Y. Xiong, Y. Xie, F. Zhang, E. Ou, Z. Jiang, L. Ke, D. Hu, W. Xu, Reduced graphene oxide/hydroxylated styrene–butadiene–styrene tri-block copolymer electroconductive nanocomposites: preparation and properties, Mater. Sci. Eng. B 177 (2012) 1163–1169.
[97] H.K. Choi, Y. Oh, H. Jung, H. Hong, B.-C. Ku, N.-H. You, Y.-K. Kim, E.S. Shin, J. Yu, Influences of carboxyl functionalization of intercalators on exfoliation of graphite oxide: a molecular dynamics simulation, Phys. Chem. Chem. Phys. 20 (2018) 28616–28622.
[98] P. Song, Z. Xu, Y. Wu, Q. Cheng, Q. Guo, H. Wang, Super-tough artificial nacre based on graphene oxide via synergistic interface interactions of π-π stacking and hydrogen bonding, Carbon 111 (2017) 807–812.
[99] M. He, R. Zhang, K. Zhang, Y. Liu, Y. Su, Z. Jiang, Reduced graphene oxide aerogel membranes fabricated through hydrogen bond mediation for highly efficient oil/water separation, J. Mater. Chem. A 7 (2019) 11468–11477.
[100] E.-Y. Choi, T.H. Han, J. Hong, J.E. Kim, S.H. Lee, H.W. Kim, S.O. Kim, Noncovalent functionalization of graphene with end-functional polymers, J. Mater. Chem. 20 (2010) 1907–1912.
[101] P. Bhunia, E. Hwang, M. Min, J. Lee, S. Seo, S. Some, H. Lee, A non-volatile memory device consisting of graphene oxide covalently functionalized with ionic liquid, Chem. Commun. 48 (2012) 913–915.
[102] X. Wang, G. Shi, An introduction to the chemistry of graphene, Phys. Chem. Chem. Phys. 17 (2015) 28484–28504.
[103] H. Cui, Z. Jin, D. Zheng, W. Tang, Y. Li, Y. Yun, T.Y. Lo, F. Xing, Effect of carbon fibers grafted with carbon nanotubes on mechanical properties of cement-based composites, Construct. Build Mater. 181 (2018) 713–720.

[104] B. Liu, Z. Liu, X. Wang, G. Zhang, S. Long, J. Yang, Interfacial shear strength of carbon fiber reinforced polyphenylene sulfide measured by the microbond test, Polym. Test. 32 (2013) 724–730.

[105] M. Lu, H. Xiao, M. Liu, X. Li, H. Li, L. Sun, Improved interfacial strength of SiO_2 coated carbon fiber in cement matrix, Cem. Concr. Compos. 91 (2018) 21–28.

[106] L.I. Nasibulina, I.V. Anoshkin, S.D. Shandakov, A.G. Nasibulin, A. Cwirzen, P.R. Mudimela, K. Habermehl-Cwirzen, J.E. Malm, T.S. Koltsova, Y. Tian, Direct synthesis of carbon nanofibers on cement particles, Transp. Res. Rec. 2142 (2010) 96–101.

[107] S. Sun, X. Yu, B. Han, J. Ou, In situ growth of carbon nanotubes/carbon nanofibers on cement/mineral admixture particles: a review, Construct. Build Mater. 49 (2013) 835–840.

[108] A.G. Nasibulin, S.D. Shandakov, L.I. Nasibulina, A. Cwirzen, P.R. Mudimela, K. Habermehl-Cwirzen, D.A. Grishin, Y.V. Gavrilov, J.E. Malm, U. Tapper, A novel cement-based hybrid material, New J. Phys. 11 (2009), 023013.

[109] P.R. Mudimela, L.I. Nasibulina, A.G. Nasibulin, A. Cwirzen, M. Valkeapää, K. Habermehl-Cwirzen, J.E. Malm, M.J. Karppinen, V. Penttala, T.S. Koltsova, Synthesis of carbon nanotubes and nanofibers on silica and cement matrix materials, J. Nanomater. 2009 (2009).

[110] P. Ludvig, J.M. Calixto, L.O. Ladeira, I.C. Gaspar, Using converter dust to produce low cost cementitious composites by in situ carbon nanotube and nanofiber synthesis, Materials 4 (2011) 575–584.

[111] P. Ludvig, L. Ladeira, J. Calixto, I. Gaspar, V. Melo, In-Situ Synthesis of Multiwall Carbon Nanotubes on Portland Cement Clinker, 2009.

[112] O.M. Dunens, K.J. MacKenzie, A.T. Harris, Synthesis of multi-walled carbon nanotubes on fly ash derived catalysts, Environ. Sci. Technol. 43 (2009) 7889–7894.

[113] X. Zhang, Z. Liu, Recent advances in microwave initiated synthesis of nanocarbon materials, Nanoscale 4 (2012) 707–714.

[114] X. Zhang, S.K. Manohar, Bulk synthesis of polypyrrole nanofibers by a seeding approach, J. Am. Chem. Soc. 126 (2004) 12714–12715.

[115] Z. Liu, J. Wang, V. Kushvaha, S. Poyraz, H. Tippur, S. Park, M. Kim, Y. Liu, J. Bar, H. Chen, Poptube approach for ultrafast carbon nanotube growth, Chem. Commun. 47 (2011) 9912–9914.

[116] N. Farzadnia, S.H. Bahmani, A. Asadi, S. Hosseini, Mechanical and microstructural properties of cement pastes with rice husk ash coated with carbon nanofibers using a natural polymer binder, Construct. Build Mater. 175 (2018) 691–704.

[117] M.R. Malekbala, M.A. Khan, S. Hosseini, L.C. Abdullah, T.S. Choong, Adsorption/desorption of cationic dye on surfactant modified mesoporous carbon coated monolith: equilibrium, kinetic and thermodynamic studies, J. Ind. Eng. Chem. 21 (2015) 369–377.

[118] V. Anggraini, A. Asadi, N. Farzadnia, H. Jahangirian, B.B. Huat, Reinforcement benefits of nanomodified coir fiber in lime-treated marine clay, J. Mater. Civ. Eng. 28 (2016), 06016005.

[119] V. Anggraini, A. Asadi, N. Farzadnia, H. Jahangirian, B. Huat, Effects of coir fibres modified with Ca (OH) 2 and Mg $(OH)_2$ nanoparticles on mechanical properties of lime-treated marine clay, Geosynth. Int. 23 (2016) 206–218.

[120] S. Bordoloi, D. Patwa, R. Hussain, A. Garg, S. Sreedeep, Nano-particle coated natural fiber impregnated soil as a sustainable reinforcement material, in: Urbanization Challenges in Emerging Economies: Resilience and Sustainability of Infrastructure, American Society of Civil Engineers Reston, VA, 2018, pp. 435–444.

[121] Z.S. Metaxa, M.S. Konsta-Gdoutos, S.P. Shah, Carbon nanofiber–reinforced cement-based materials, Transp. Res. Rec. 2142 (2010) 114–118.

[122] B. Han, X. Yu, J. Ou, Multifunctional and smart carbon nanotube reinforced cement-based materials, in: Nanotechnology in Civil Infrastructure, Springer, 2011, pp. 1–47.
[123] G. Collodetti, P.J. Gleize, P.J. Monteiro, Exploring the potential of siloxane surface modified nano-SiO_2 to improve the Portland cement pastes hydration properties, Construct. Build Mater. 54 (2014) 99–105.
[124] J.-L. Luna-Xavier, A. Guyot, E. Bourgeat-Lami, Synthesis and characterization of silica/poly (methyl methacrylate) nanocomposite latex particles through emulsion polymerization using a cationic azo initiator, J. Colloid Interface Sci. 250 (2002) 82–92.
[125] Y. Xu, D. Chung, Silane-treated carbon fiber for reinforcing cement, Carbon 39 (2001) 1995–2001.
[126] G.Y. Li, P.M. Wang, X. Zhao, Mechanical behavior and microstructure of cement composites incorporating surface-treated multi-walled carbon nanotubes, Carbon 43 (2005) 1239–1245.
[127] L.I. Nasibulina, I.V. Anoshkin, A.V. Semencha, O.V. Tolochko, J.E. Malm, M.J. Karppinen, A.G. Nasibulin, E.I. Kauppinen, Carbon nanofiber/clinker hybrid material as a highly efficient modificator of mortar mechanical properties, Mater. Phys. Mech. 13 (2012) 77–84.
[128] P. Hlaváček, V. Šmilauer, Fracture properties of cementitious composites reinforced with carbon nanofibers/nanotubes, Eng Mech. 211 (2012) 391–397.

CHAPTER 3

Cement-based nano-engineered materials for eco-efficiency

Konstantin Sobolev[a,b], Marina I. Kozhukhova[a,c], and Reed T. Heintzkill[b]
[a]Department of Civil & Environmental Engineering, College of Engineering and Applied Science, University of Wisconsin–Milwaukee, Milwaukee, WI, United States
[b]Department of Materials Science & Engineering, College of Engineering and Applied Science, University of Wisconsin–Milwaukee, Milwaukee, WI, United States
[c]Ramboll, Milwaukee, WI, United States

3.1 Introduction

With ever-growing global demand for infrastructure development and renewal, world production of portland cement (PC) has nearly tripled in the last 20 years—from 1.6B metric tons in 1999 to an estimated 4.2B in 2019 [1,2]. This trend is the most significant factor affecting the updating of manufacturing facilities and the overall technological development of the cement and concrete industry [1–3]. PC manufacturing requires great energy inputs in the calcining of raw materials through pyro-processing, and in addition to the fuel combustion cost, this process releases carbon dioxide at a 1-to-1 M rate from its calcium carbonate raw starting material. In addition to contributing an estimated 5% of annual global CO_2 emissions, cement production also emits a number of other pollutants such as NO_x and SO_x [3,4], the greenhouse effects of which are well known and documented. The mere scale of worldwide production of cement and concrete-adjacent materials makes this sector a prime focus area for ecological stewardship and sustainability, while these materials' resilience and expansive adaptability also make them uniquely apt candidates for substantial material improvement through nano-engineering.

It is crucial for the technologies that drive efficiency in cement and concrete manufacturing to advance in parity with the industry's inevitable expansion under modern global market demands. To allow concrete production to continue to expand as it has—unchecked, and not significantly improved technologically—is to invite market unsustainability and ecological impact. *There is an urgent need for a paradigm shift*; for the development of new and sustainable cement, technology may minimize the environmental

Nanotechnology for Civil Infrastructure
https://doi.org/10.1016/B978-0-12-817832-4.00002-X

damage caused by cement production while offering new products that have better performance, consume less energy, use fewer raw materials, and emit fewer harmful emissions [5–20].

Nanoscience has changed the face of countless sectors of industry over the past 50 years. Emergent technologies such as semiconductors and microprocessing, aviation and aerospace, or polymer consumer products are examples of areas that have undergone multiple revolutions, sometimes year after year.

The field of construction and building materials is a less obvious benefactor of nanotech-based improvements because such gains are less consumer-facing, and existing technology is efficient and established enough that there is less of an imperative for immediate and rapid change [21–29]. Nanotechnology, nanomaterials, and nano-engineering represent a ray of light shining through the cloud of unsustainable legacy materials and manufacturing practices. To be implemented rapidly and effectively, these oncoming technologies must be understood, embraced, and executed.

This chapter provides an overview of contemporary ongoing research, postulated future developments, and general zones of innovation in which nanotechnology may be implemented into cement and concrete materials to enhance ecologic impacts and maximize long-term sustainability.

3.2 Nanotechnology in cement-based materials

While precise definitions and explanations of various aspects of nanotechnology have undoubtedly been covered elsewhere in this book, it is important to clarify the jargon and define the appropriate terms for nanomaterials as they apply to cement and concrete-based systems. In general, nanomaterials may be defined as any material substance in which one discrete unit or dimension exists specifically on the order of magnitude of the nanometer (1–100 nm) [21,29]; the breadth of nanomaterials may be further broken down into three primary categories:

Nanoparticles: Any discrete unit or *particle* in which the longest and shortest dimensional axes are relatively near on a nanometric scale. Nanoparticles in cements may be further delineated as nanoscale (1–20 nm) inorganic particles surrounded by a distinct interfacial layer that acts as a separation from one NP to another. Surface modification with organic leaving groups and polymeric monolayers between mono- or polycrystalline and even amorphous inorganics becomes of significant interest in many areas of concrete nanotechnology [21].

Nanofibers and nanotubes: Nano-objects in which two of the three dimensions are not significantly different in their nanometric scale, while the scale of the third dimension is at least three times greater than the other two [21,22]. Nanotubes are generally characterized by a 2D-monomolecular lattice akin to graphene "wrapped" into a continuous tubular shape. Such nanotubes may consist of multiple such layers (e.g., "multi-walled nanotubes") and may have "hollow-body" and even capillary properties. In contrast, nanofibers tend to be more akin to traditional mono- or polycrystalline nanoparticles that extend significantly in one direction.

Nanofilms, layering, and surface-science: Single-phase layers with thickness on the nanoscale, a nanocomposite materials are defined as solid systems composed of two or more phase-separated components, at least one of which has nanoscale properties, representing a nanocomposite [22,27,29]. For example, embedding alumina nanofiber into a resin matrix would create a single-phase nanocomposite, where blending cementitious nanoparticles and nanotubes in a PC matrix would be considered a dual-phase nanocomposite.

While nanofibers/tubes and nanofilms represent some burgeoning areas of exciting development in concrete and materials science, this chapter focuses mostly on nanoparticles: those currently in use and under development, and technologies aimed at incorporating such materials.

3.3 Effects of nanoparticles on the mechanical performance and durability of concrete

Submicron supplementary cementitious materials such as silica fume (SF) and metakaolin have been effectively used to improve the long-term mechanical response and durability of cement-based materials [7,18]. At the nanoscale, nanoparticles of SiO_2 ("nanosilica") proved to be effective in cement composites to modify rheological behavior, to enhance the reactivity of supplementary cementitious materials, and to improve strength and durability [22,25–29] (Fig. 3.1).

Due to pozzolanic properties, like Class F fly ash, nanosilica reacts with calcium hydroxide, creating more C—S—H, reducing the permeability of the hydrated cement matrix and increasing the strength and durability of concrete [30,31]. Belkowitz and Armentrout [32] also proved that aside from improved strength, the addition of nanosilica tends to increase the electrical resistivity of concrete from 30% to 700% while increasing the critical size of the nanosilica particle from 5 to 46 nm. Such considerable

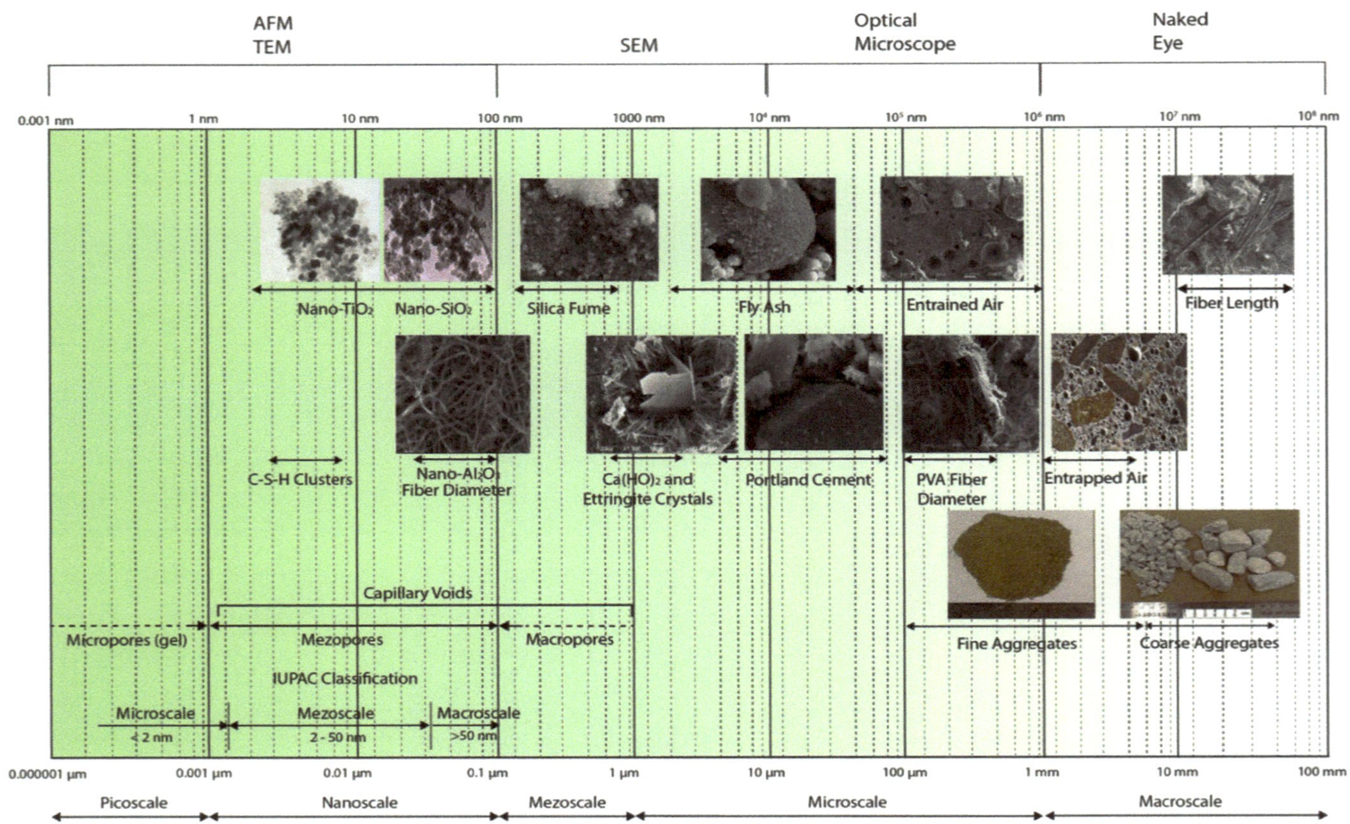

Fig. 3.1 The scale ranges related to concrete and nanoparticles [7].

improvement in electrical resistivity was explained by the formation of a denser matrix, reducing the permeability of the concrete samples.

The effects of the addition of colloidal silica to the hydration process of PC minerals C_3S (alite, Ca_3SiO_5) [6] and β-C_2S (belite, β-Ca_2SiO_4) were investigated by Björnström et al. [33]. Nanoparticles were shown to accelerate the formation of C—S—H gel within the first 24 h, especially when used at higher dosages of up to 2%. Studies by Ye et al. then identified the development of $Ca(OH)_2$ phase in pastes and concrete with SF and nanosilica additions [34]. Nanosilica increased the viscosity and reduced the setting time of these pastes. In aggregate-included concrete samples, the addition of nanosilica reduced the quantity and the size of $Ca(OH)_2$ phase crystals formed in the interfacial transition zone (ITZ), and early-age mechanical performance improved over specimens produced with SF. The addition of nanosilica, even at small dosages (3%), reduced the amount of CH formed at the aggregate's interface zone and reduced the size of CH crystals, thereby improving the ITZ to levels SF could not achieve.

Studies by Porro et al. investigated the effect of the size of SiO_2 nanoparticles used at different dosages on the performance of PC pastes [35]. Compressive strength of the pastes increased with reduction of particle size. This strength improvement was attributed to the formation of larger silicate chains within the C—S—H gel in mixtures containing SiO_2 nanoparticles.

Collepardi et al. reported that nanoparticles enable the formation of cohesive concrete microstructure with reduced segregation and bleeding [31] by producing a self-consolidating mix with low heat of hydration (HoH), using up to 2% of colloidal silica as a pozzolan and a viscosity-modifying agent.

Flores-Vivian et al. [28] studied the effects of sol–gel synthesized nanosilica on the strength of PC mortars. The addition of a superplasticizer and an effective dispersion procedure were employed to distribute these nanomaterials within the cementitious matrix and to enhance the mechanical performance of mortars. It was further demonstrated by Gaitero et al. [36] that the addition of nanosilica to cementitious systems can improve the strength by controlling the structure of C—S—H. Nanosilica with particle sizes between 5 and 50 nm was used in the field project as a viscosity-modifying agent at a dosage of 1% of cementitious materials [26,37].

Nanosilica is by far the most broadly studied, understood, and utilized nanomaterial in modern cement and concrete technology. Nanosilica can be used as a viscosity-modifying agent even in high-density, high-strength cementitious grout [37]. The grout mixture was designed to match the properties of parent in-situ rock in respect to compressive strength (more than 70 MPa), density (at least 2600 kg/m^3), and ultrasonic pulse velocity (minimum 3.65 km/s). The maximum temperature developed within the hardening grout was set to be less than 82°C. These properties were required after 30–50 days from the grout placement to meet the project schedule. Besides nanosilica (used at 0.33% of cementitious materials), the developed grout used 725 kg/m^3 of Type I/II PC (as per as ASTM C105), 128 kg/m^3 of SF (low-carbon byproduct of zirconia production), 1089 kg/m^3 of hematite sand (with density of 5.02 and water absorption of 0.1%), and 380 kg/m^3 of silica sand (No. 20, with density 2.62 and water absorption of 0.2%). In order to keep a relatively low W/C = 0.4 and achieve high fluidity of the grout, a high-range water-reducing agent (SP, polycarboxylate type) was incorporated at a dosage of 0.28%. The defoaming agent was used to eliminate the entrapped air within the mixture and so to improve the density and the ultrasound propagation. The field-placed grouts had a specific density of 2.6–2.75, ultrasonic pulse velocity of 4.2–4.6 km/s, and unconfined compressive strength of 70–110 MPa. These results met and exceeded the requirements set for the project.

Beneficial effects of nano-Al_2O_3 (AlF) used in small quantities (0.1 … 0.25%) and combined with 1% of SF and metakaolin (MK) were reported and considerable strength improvement of cement systems was obtained, as indicated in Fig. 3.2 [9].

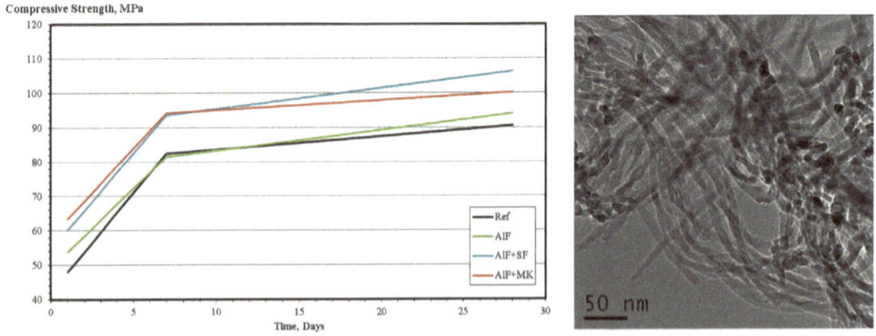

Fig. 3.2 Compressive strength of cement-based composites with nano-Al_2O_3 (AlF, right) [9].

Based on the available data, the beneficial action of the nanoparticles on the microstructure and the performance of cement-based materials can be explained by the following factors [22,27]:

- Well-dispersed nanoparticles increase the viscosity of the liquid phase, helping to suspend the cement grains and aggregates and improving the segregation resistance and workability of the system;
- Nanoparticles fill the voids between cement grains, resulting in the immobilization of "free" water ("filler" effect);
- Well-dispersed nanoparticles act as centers of crystallization of cement hydrates, thereby accelerating the hydration;
- Nanoparticles favor the formation of small-sized crystals (such as $Ca(OH)_2$ and AF_m) and small-sized uniform clusters of C—S—H;
- Nanosilica participates in the pozzolanic reactions, resulting in the consumption of $Ca(OH)_2$ and formation of an "additional" C—S—H;
- Nanoparticles improve the structure of the aggregates' contact zone, resulting in a better bond between aggregates and cement paste; and
- Crack arrest and interlocking effects between the slip planes provided by nanoparticles improve the toughness, shear, tensile, and flexural strength of cement-based materials.

3.4 Sol-gel synthesis of nanosilica

The relatively small quantities, less than 1% (by weight of cement), of nanosized materials are sufficient to improve the performance of nanocomposites [9,22,26–30]; yet, the commercial success of nanomaterials depends on the ability to manufacture these materials in large quantities, and at a reasonable cost relative to the overall effect of the nanoproduct. Nanomaterial technologies, which could lead to the industrial outputs, involve plasma arcing, flame pyrolysis, chemical vapor deposition, electrodeposition, sol–gel synthesis, mechanical attrition, and the use of natural nanosystems [21]. Among chemical technologies, sol–gel synthesis is one of the widely used "bottom-up" production methods for nanosized materials such as nanosilica. The process involves the formation of a colloidal suspension (sol) and gelation of the sol to form a network in a continuous liquid phase (gel). Usually, trimethylethoxysilane or tetraethoxysilane is applied as a precursor for synthesizing nanosilica. There are a number of parameters that affect the process, including pH, temperature, the concentration of reagents, H_2O/Si molar ratio (between 7 and 25), and type of catalyst [28]. When precisely executed, this process is capable of producing perfectly spherical nanoparticles of

SiO_2 within the size range of 1–100 nm. The chemical reaction of nanosilica synthesis can be summarized as

$$n\mathrm{Si(OC_2H_5)_4} + 2n\mathrm{H_2O} \xrightarrow[\mathrm{NH_3}]{\mathrm{C_2H_5OH}} n\mathrm{SiO_2} + 4n\mathrm{C_2H_5OH}$$

The details on the "bottom-up" synthesis of nanosilica particles and the effect of this material on the performance of cement systems were reported [28]. Nanoparticles of SiO_2 (with the size range of 5–100 nm) were synthesized using the sol-gel method. Tetraethoxysilane was used as a precursor and the reaction was realized in a base or acid reaction media with ammonia as a catalyst (ammonia solution at pH = 9). The specimens of nanosilica were obtained at different experimental conditions (ethanol-to-TEOS molar ratios 6 or 24 and water-to-TEOS ratios 6 or 24), as presented in Table 3.1.

According to the results of X-ray diffraction, the obtained nanosilica is a highly amorphous material with a predominant crystallite size of 1–2.5 nm [28]. Tunneling electron microscopy (TEM) was used to characterize the morphology and particle size distribution of nanosilica (Table 3.1 and Fig. 3.3). The obtained nanosilica particles are represented by highly agglomerated xerogel clusters with a size of 0.5–10 μm. The performance of nanosilica-based mortars (at nanosilica dosage of 0.25% by the weight of the binder, water-to-cement ratio, W/C of 0.3 and sand-to-cement ratio, S/C of 1) was compared with the properties of two reference mixtures: plain mortar (with W/C of 0.3 and sand-to-cement ratio of 1) and

Table 3.1 Design and properties of nanosilica [28].

Specimen type[a]	Molar ratio TEOS/Ethanol/H_2O	Reaction time, h	Particle size (TEM), nm	Surface area (BET), m^2/kg
1B3	1/24/6	3	15–65	116,000
2B3	1/6/6	3	30, 60–70	145,000
3B3	1/6/24	3	15–20	133,000
4B3	1/24/24	3	5	163,000
1A3	1/24/6	3	5	510,100
2A3	1/6/6	3	<10	263,500
3A3	1/6/24	3	<10, 17	337,100
4A3	1/24/24	3	5	382,200

[a]Sample coding:

ABC —
- First number denotes molar ratio combination as per as experimental matrix
- Last number corresponds to the reaction time in hours
- Letter – denotes reaction media: A- for acid and B- for base

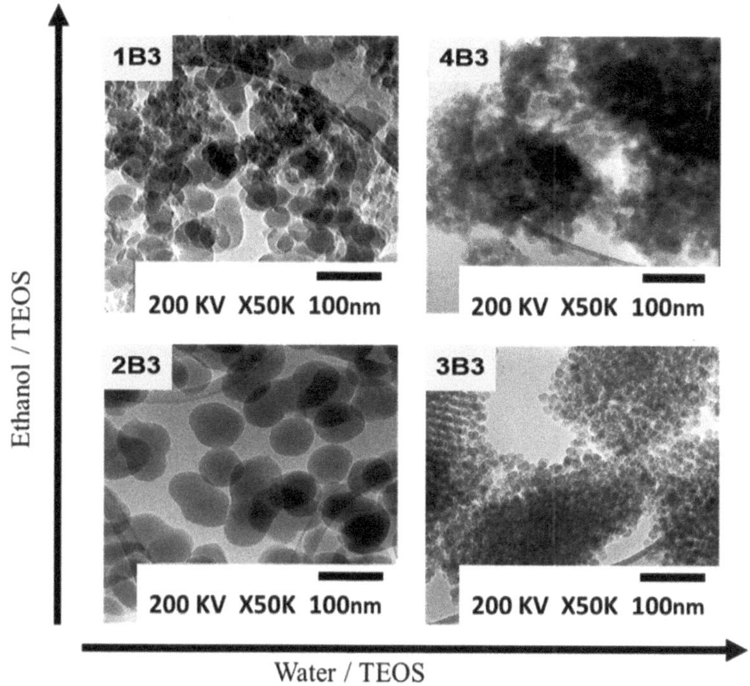

Fig. 3.3 Morphology and particle size distribution of nanosilica (TEM) synthesized by the sol–gel method using various molar ratios of TEOS/ethanol/H_2O at pH = 9 [28].

superplasticized mortar (at SP dosage of 0.1%). The best nanosilica products with particle size ranging from 5 to 20 nm are synthesized at the highest molar concentrations of water (specimens 3B and 4B). The performance of nanosilica depends on the conditions of synthesis (i.e., molar ratios of the reagents, type of the reaction media, pH, and the duration of the reaction) [27,28].

According to the results of the experiment, the strength of superplasticized cement mortars with nanosilica was improved. The addition of 0.25% of nanosilica (specimen 4B3, manufactured at TEOS/ethanol/H_2O ratio of 1:24:24 using base medium) to superplasticized mortar provided the 1-day compressive strength of 63.9 MPa, an increase of about 20% versus plain cement mortar. This value also represents a 16% improvement in compressive strength versus superplasticized cement mortar. However, only a slight improvement in compressive strength, about 4% (vs. strength of superplasticized mortar), was observed at the 28-day age, reaching 95.9 MPa. The highest compressive strength was achieved by the superplasticized mortar with 0.25% of nanosilica (specimen 3A3, manufactured at TEOS/ethanol/H_2O ratio of 1:6:24 using base medium), which exhibited compressive strength of 75.1, 80.3, and 96.5 MPa at the age of 3, 7, and 28 days, respectively. Application of SF and commercial nanosilica (CB8) in superplasticized mortars demonstrated similar behavior; however, the increase in the 1-day compressive strength was only 8% (vs. the strength of superplasticized mortar). It can be expected that the application of wider range of particle sizes and better dispersion of nanosilica can further improve the performance of engineered nanoparticles [28].

3.5 Tuning nanosilica for optimal performance

Thomas et al. demonstrated that the addition of relatively small amounts of specially synthesized calcium silicate hydrate (C—S—H) to PC and pure tricalcium silicate (C_3S) pastes causes a significant acceleration of the early hydration kinetics (Fig. 3.4) [38]. Here, C—S—H provides nucleation sites for the precipitation of hydration products out of the pore solution, such as the C—S—H gel, which is the main binding phase in cement paste. The experiments indicated that the C—S—H seed causes hydration products to form within the pore space between the cement or C_3S particles, whereas during common hydration nucleation occurs only on or near the particle surfaces. The presence of two simultaneous hydration processes in seeded pastes (one initiated at the particle surfaces and one in the pore space) results in the development of a second-rate peak or shoulder in the calorimetry data that increases in proportion to the amount of C—S—H seed that is added. The addition of C—S—H seed results in permanent changes to the hydrated matrix, including a denser microstructure with less capillary porosity.

The mechanical attrition or mechanochemical activation of powder particles is an effective method capable of synthesizing an extensive range of nanomaterials at an industrial scale [21,39]. With mechanical attrition, the intermixing of individual components, amorphization, and the formation of new compounds due to solid-state reactions are realized. The synthesis process is governed by both the mechanical alloying and the incorporation of lattice defects into the crystal structure. It was demonstrated that the mechanochemical activation is also a very effective method to attach the organic functional groups to the surface of inorganic powders, such as PC. Activation unit can be used as a low-cost reactor for controlled transformations in aqueous media, such as required for the synthesis of nanomaterials with improved mechanical performance [30]. The nanosized

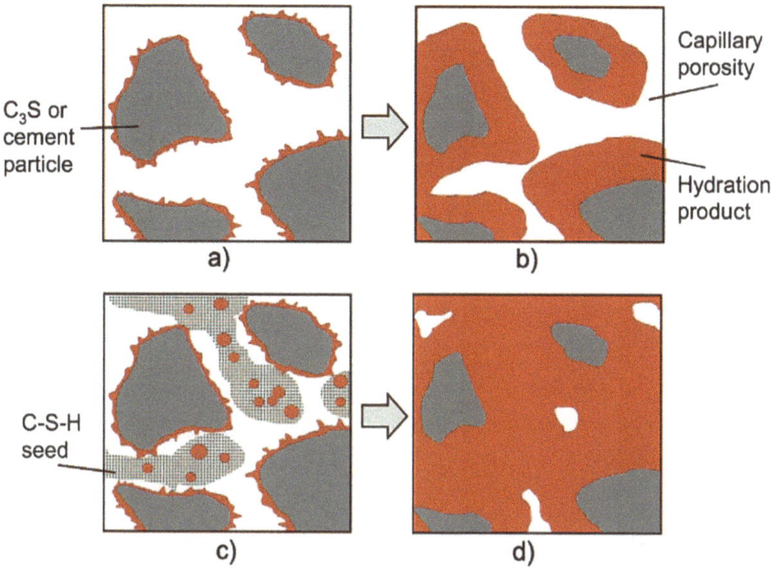

Fig. 3.4 Schematic of the hydration process of cement paste with C—S—H seed. (A) Normal unseeded paste after mixing; the hydration product (C—S—H gel) nucleates on particle surfaces and (B) begins to grow out into the pore space, leaving significant capillary porosity; (C) paste with C—S—H seed after mixing; (D) the hydration product nucleates both on the particle surfaces and within the regions of C—S—H seed, increasing the rate of early hydration, and low capillary porosity. *(From J.J. Thomas, et al., Influence of nucleation seeding on the hydration mechanisms of tricalcium silicate and cement, J. Phys. Chem. C 113 (2009) 4327–4334, Courtesy of the author.)*

cementitious additives can be produced by ultramilling using mechanochemically induced topochemical reactions or by self-assembly from precursors [22,30,40]. The first method can be attractive for large-scale industrial manufacturing of a nanoseed material with enhanced C—S—H structure. The feasibility of the top-down synthesis of C—S—H nanoseed clusters from hydrating cement and fly ash precursors was demonstrated [30,40]. The developed cementitious materials were produced in the form of a slurry using a nano-activation unit. The composite with nanoseeds demonstrated superior performance in respect to early and ultimate compressive strength [30,40]. It was reported that the use of nanosized cementitious additives, including low-cost activated cement products obtained by the top-down colloidal milling, can significantly (by 50%) improve the early strength of cement-based materials, as shown in Fig. 3.5 [30].

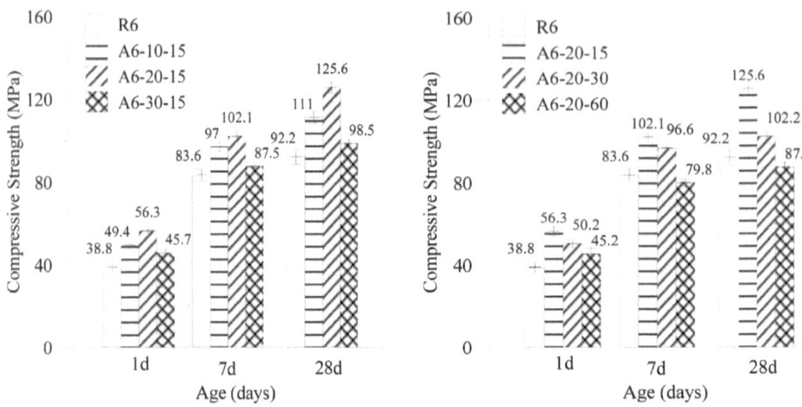

Fig. 3.5 Compressive strength in proposed nanoseed based composites (A6) versus reference (R6) mixture [30].

3.6 Nanosilica harvested from nature

Despite the attractive performance, the application of nanoparticles in concrete remains limited because of the high cost associated with the production of nanomaterials. The quest for nanomaterials lies in exploring the pathways for relatively inexpensive bulk production of nanoparticles for concrete. Low-cost silica nanoparticles can be obtained from natural sources such as hydrothermal solutions. Steam or hot water erupting from the earth's surface due to geothermal activities has been historically used for bathing and processing of food. Currently, high-temperature steam and water are used for power generation and also as a supply of heat. In a geothermal power plant, hot water rises to the surface, evaporates, and the steam is redirected to spin turbines in the generation of electrical energy. The steam is condensed into water in a cooling tower, where minerals (including nanosilica) must be recovered and potentially utilized for various applications. The minerals found in geothermal power plant water streams include zinc, silica, lithium, manganese, boron, lead, silver, antimony, and strontium, as quantified in Table 3.2. In volcanic areas, orthosilicic acid is formed in hydrothermal solutions due to the dissolution of aluminosilicate minerals of rocks under elevated pressures and temperatures. Concentrated solutions travel to the surface, where the pressure and temperature are lower, and due to hydrolysis and polycondensation, spherical silica nanoparticles are formed. In geothermal plants, silica particles cause clogging of pipes, and, therefore, nanosilica extraction increases the efficiency of the plant [29]. Here, the use of

Table 3.2 Composition of hydrothermal solutions [29].

Component	Na^{+}	K^{+}	Li^{+}	Ca^{2+}	Mg^{2+}	Fe^{x+}	Al^{3+}	Cl^{-}	SO_4^{2-}	HCO_3^{-}	CO_3^{2-}	H_3BO_3	SiO_2
Concentration, mg/L	282	48.1	1.5	2.8	4.7	<0.1	<0.1	251.8	220.9	45.2	61.8	91.8	780

hydrothermal solutions as a relatively cheap natural precursor for nanosilica production can reduce the costs associated with the production and application of nanomaterials.

In the reported research [29], SiO_2 particles were harvested through membrane ultrafiltration and tested in PC mortars. Nanodispersed powders can be obtained from the solutions by removing the excess water by filtering through the membrane and by subsequent cryo-chemical vacuum sublimation drying using liquid nitrogen. The lab-grade products (MB, TB, and N_2) were compared with commercially available nanosilica admixture (Cembinder-8, derived from geothermal solutions) which is often used as a reference nanosilica. The performance of three types of nanoparticles produced by ultrafiltration from the geothermal solutions in the form of a gel (specimen MB) and dry powder specimens (TB and N_2) obtained by subsequent sublimation drying using liquid nitrogen were reported (Table 3.3).

The effects of hydrothermal nanosilica on the hydration process were analyzed by monitoring the HoH with an isothermal calorimeter (Fig. 3.6). The isothermal calorimetry response was used to determine the

Table 3.3 Characterization of silica nanoparticles.

	Material	Solid concentration, %	BET surface area, m^2/g
Reference	CB8	52	61
Sol	MB	35	352
Dry	TB	100	43
	N_2	100	228

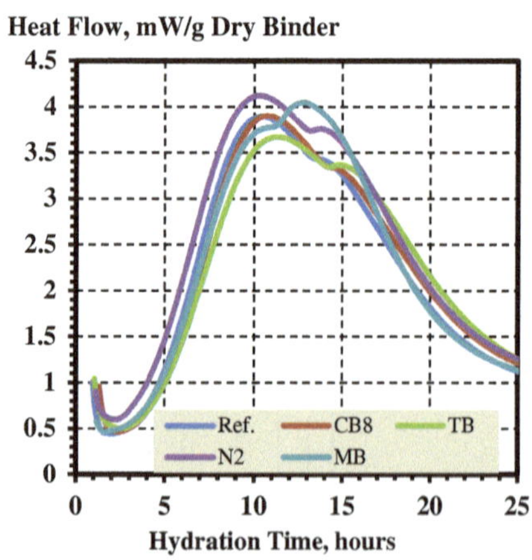

Fig. 3.6 The shift of C_3S and C_3A peaks due to the addition of nanosilica [29].

setting times and heat released during the hydration. The first (C_3S) and second (C_3A) peaks for the reference system and mortars with CB8 were observed at 10.5 and 14 h after mixing with the C_3A characterized by lower heat flow. The addition of powder nanosilica products (such as N_2 and TB) had a considerable effect on the HoH. The N_2 nanoparticles had a profound increase in the HoH. Increased rates of C_3S hydration can be correlated with the BET surface area, as a higher surface area can lead to the acceleration of C—S—H gel formation due to the seed effect. With the addition of N_2 particles, the first peak corresponding to C_3S hydration was observed after 10 h followed by the second peak at 14 h. For compositions with N_2 nanoparticles, the rate of hydration heat was about 5% higher compared with the reference and CB8-based mortars, as indicated in Fig. 3.6. For TB mortars, the first peak (C_3S) was observed after 11.3 h, and the second peak (corresponding to C_3A) was identified at 15 h at the rate of hydration of 5% lower compared with the reference and CB8-based mortars. With the addition of nanosilica, the hydration of C_3S was accelerated due to a heterogeneous nucleation effect. Heterogeneous nucleation is typically the dominant acceleration mechanism when well-dispersed seed components with a size smaller than cement particles are incorporated.

The C—S—H is the main product of cement hydration and a primary binding phase in PC. When superfine particles of higher surface area (such as nanoparticles) are added, a denser matrix is created due to the formation of uniform, well-distributed C—S—H gel structure. As a result, the mechanical performance and durability in cement systems with nanosilica are improved. The increased hydration rates of C_3S can be correlated to the BET surface area where materials with higher surface area accelerate the formation of C—S—H gel. The compressive and splitting tensile strengths of mortars with nanosilica at different curing ages are reported in Fig. 3.7. At the age of 28 days, the specimens containing nanosilica in gel form (MB) had the highest compressive strength of 110 MPa. There was an increase in the 1-day strength of mortars with gel form (MB) and dry powder form (N_2) nananosilica, which corresponds to about 12% and 19% improvement compared with the reference mortar. The splitting tensile strength of the investigated mortars followed similar trends and can be correlated to compressive strength, especially at the age of 3 days and afterward. It can be observed that the use of some nanosilica products obtained from hydrothermal solutions at a very small dosage of 0.25% (by weight of cementitious materials) can provide a considerable (up to 18%) strength enhancement of cement-based materials.

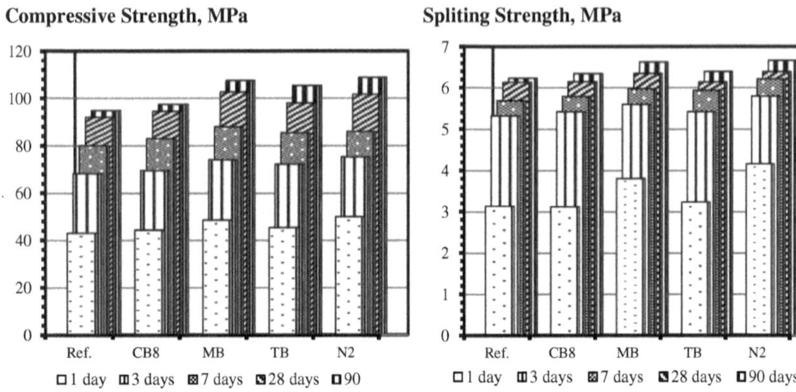

Fig. 3.7 Compressive and splitting strengths of mortars made with nanosilica particles [29].

Based on the results of the reported investigation, it can be concluded that the use of nanosilica particles obtained from the hydrothermal solutions can improve the performance of PC systems.

3.7 Natural and plant-based nanomaterials

Natural biomaterials such as woods and plant fibers have been used in construction materials since the dawn of humanity, and byproducts (e.g., cellulose, lignin) from processing or burning of organics are commonly used materials. Rice is the world's greatest agricultural product, and as such a tremendous amount of rice husks are used as combustible fuel, the ash of which happens to have desirable chemical makeup for use as an SCM.

The authors [41] reported on a belite cement production using nanosilica extracted from a pure rice husk ash calcined in the laboratory (RHA) and ash from an impure industrial rice husk waste (BRHA). The crystalline size of the nanosilica extracted from both sources was in the range of 20–50 nm and possessed high surface areas as shown in Table 3.4. The nanosilica extracted from RHA was highly reactive due to its high pore volume and low activation energy of dehydration. Extraction of nanosilica from BRHA transformed the crystalline impurities (principally cristobalite) to amorphous SiO_2 with a high surface area, the activation energy of dehydration was relatively high compared with RHA.

Belite cement was prepared from the nanosilicas extracted from pure rice husk ash by reaction with laboratory-grade $Ca(NO_3)_2$ and $CaCO_3$ at a Ca/Si ratio of 2:1 (Table 3.4). The dry ingredients were fired in air for

Table 3.4 Properties and thermal activation energy of nanosilica [41].

Parameter	Extracted nanosilica	
	RHA	BRHA
Surface area, m^2/g	211.3	312.4
Total pore volume, cm^3/g	1.114	1.094
Average pore diameter, Å	210.8	140.1
Activation energy, kJ/mol		
$\alpha=1\%$	138.8	256.0
$\alpha=2\%$	135.3	272.0
$\alpha=3\%$	135.6	302.3
Average	136.6 ± 1.6	276 ± 19.2

60 min at 800, 900, 1000, and 1100°C. Structural analyses using XRD patterns of the produced belite cement confirmed the formation of indicative mineral phases in the binder. It was demonstrated that nanosilica extracted from ash was sufficiently reactive to form belite cement, since the principal phase (larnite or b-C_2S) was formed at temperatures as low as 800°C, especially with calcium nitrate as the calcium source. Thus, highly impure BRHA is quite suitable as a starting material to convert to nanosilica with good reactivity for the low-temperature production of belite cement, especially in conjunction with calcium nitrate as the calcium source. It is the management of waste generated from agriculture and adds value to the waste more efficiently [41].

3.8 Nanosilica in self-consolidating concrete

Collepardi et al. studied the performance of low-heat self-consolidating concrete (SCC) produced with nano-SiO_2. Nanosilica (on the order of 5–50 nm) was used as viscosity-modifying agent at a dosage of 1% of cementitious materials [26,31]. The difficult task of designing SCC with low-heat generation properties was achieved by applying blended blast furnace slag cement (60% BFS/PC) with low HoH to a mineral additive such as ground limestone, fly ash, or milled fly ash, thus maintaining the relatively high volume of cement paste, crucial to SCC design. Importantly, the ground fly ash (with particle size of 1–10 μm and specific surface area of 2 m^2/g) and nano-SiO_2 were used in a form of slurry. The properties of nine concrete mixtures, designed with approximately 305 kg/m^3 of blended cement, were investigated. Therefore, the PC component was limited to 120 kg/m^3. The content of ground limestone was 157 kg/m^3 and fly ash/ground fly ash was about 130 kg/m^3. A constant W/B of 0.58 was used for all mixtures and slump flow of 780–800 mm was maintained by adjusting the dosage of an acrylic polymer-based superplasticizer. In order to maintain the specified slump flow, the superplasticizer dosage was increased by approximately 0.21% per each percent of nanosilica used.

It was reported that the addition of nanosilica makes the concrete mixture more cohesive and reduces bleeding and segregation; at the same time, nano-SiO_2 had limited effect on slump loss measured within the first 30 min. The temperature in the nucleus of the investigated concrete, which was placed into an insulated 500 × 500 × 500 mm container, reached its maximum at the age of 3 days, and the temperature increase was less than 20°C. This feature of low-heat generation allows the application of developed concrete in massive structures. While nanosilica does not affect the compressive strength of concrete with different forms of fly ash, the compressive strength of concrete containing ground limestone was slightly reduced. The best performance was demonstrated by the concrete with ground fly ash, 2% nanosilica, and 1.5% of superplasticizer. This concrete had the highest compressive strength of 55 MPa at the age of 28 days and the desired behavior in fresh state: low bleeding, adequate cohesiveness, higher slump flow, and limited slump loss. It was demonstrated that concrete with fly ash, especially with ground fly ash, had lower penetration rate of both CO_2 and Cl. However, addition of nano-SiO_2 did not affect the penetration rates. Therefore, based on results of this investigation, it was concluded that nanosilica does not affect the durability of concrete [26].

Collepardi et al. [31] reported the performance of quarterly binders (4-component binders, 4xB) composed of blended cement (with 20% limestone), fly ash, SF, and nano-SiO_2 in high-performance concrete (HPC) and SCC. The concrete mixtures were designed to match the performance level of SF concrete using 4xB at a reduced SF content (i.e., option of SF replacement with nano-SiO_2). Therefore, two 4xB mixtures with 15–20 kg/m^3 of SF, 30–40 kg/m^3 of FA, and 5–7.5 kg/m^3 of nano-SiO_2 were compared with two reference mixtures containing 60 kg/m^3 of SF or FA alone (i.e., the total mineral additive content was kept constant, about 60 kg/m^3). Two types of concrete, HPC and SCC, were designed at a binder content of 395 and 425 kg/m^3, respectively. SCC mixtures had somehow different aggregate grading (richer in sand) and had a higher volume of mixing water (186 vs. 175 kg/m^3) and increased dosage of polycarboxilate superplasticizer (1.2% vs. 0.7%). All of the investigated mixtures had the same W/C = 0.44.

The results of the investigation show that 4xB-based mixtures perform similar to the "pure" SF-based concrete in terms of strength and durability. Importantly, 4xB mixtures performed well after accelerated steam curing at 80°C, demonstrating negligible strength loss at the age of 28–90 days when compared with corresponding concrete cured at room temperature.

3.9 Top-down production of "nanoseed" C—S—H from reactive powders

Development of new inorganic eco-binders is underway, based on ultrafine milling of extant reactive powders including PC, SCMs, fillers, and industrial byproducts (e.g., fly ash, spray-drier absorber ash, and SF). The high specific surface area of these mechanochemically activated nanomaterials may achieve the same beneficial enhancements to cement and concrete systems as synthesized nanosilica while obtained through sustainable high-volume sources such as waste streams and existing cementitious material production methods. Even in small additions, these top-down milled products are expected to similarly improve strength and durability of concrete by acting as reactive seed sites, influencing the precipitation and growth of C—S—H. Chemomechanical activation relies on particle size reduction through ultrafine milling in a sol incorporated with superplasticizer and may not follow explicitly the classical mechanism of C—S—H nucleation and growth models. Nanopowder sol suspensions so activated are expected to be utilized as admixture agents, significantly enhancing cement systems in small dosages, allowing for increased proportion of sustainable SCM in mixture designs, while also themselves being efficiently sourced.

Nanobinder was proposed as an extension of the two concepts: a densified system with ultrafine particles (DSP) [42,43] and modified multicomponent binder (MMCB) [5,7,10,17,18] down to the nanolevel (Figs. 3.2–3.10). In these systems, the densification of the binder is achieved with the help of ultrafine particles: SF dispersed with superplasticizer (SP) in DSP and finely ground mineral additives and SF modified by SP in MMCB; these particles are distributed to fill the gaps between cement grains. In these systems, PC component is used at its "standard" dispersion to provide the integrity and percolation of hydrating phases of composition. In contrast, the nanobinder was proposed to be designed with a submicron and nano-dispersed cement component to fill the gaps between the particles of mineral additives [21].

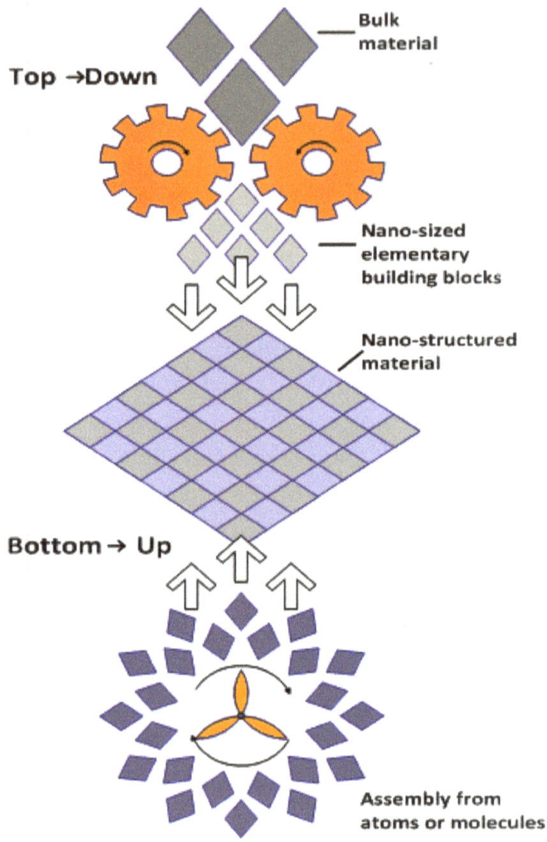

Fig. 3.8 Top-down and bottom-up approaches to nanofabrication [21].

Therefore, in nanobinder, the mineral additives (optionally, finely ground), acting as the main component, provide the structural stability of the system, and the micro- or nanosized cementitious component (which can also contain the nanosized particles other than PC) act as an inorganic glue to bind less reactive particles of mineral additives together, as shown in Fig. 3.9. Such a nanosized cementitious component can be obtained by the colloidal milling of a conventional (or specially sintered, optionally high C_2S) PC clinker (the top-down approach) or by the self-assembly using mechanochemically induced topochemical reactions (the bottom-up approach).

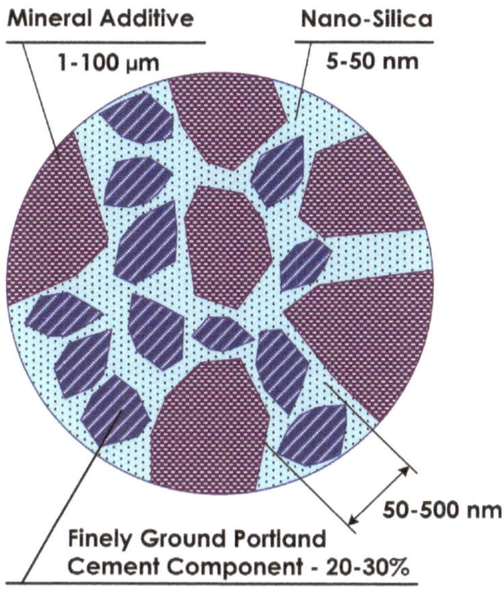

Fig. 3.9 Concept of nanobinder (nano-engineered cement) [22].

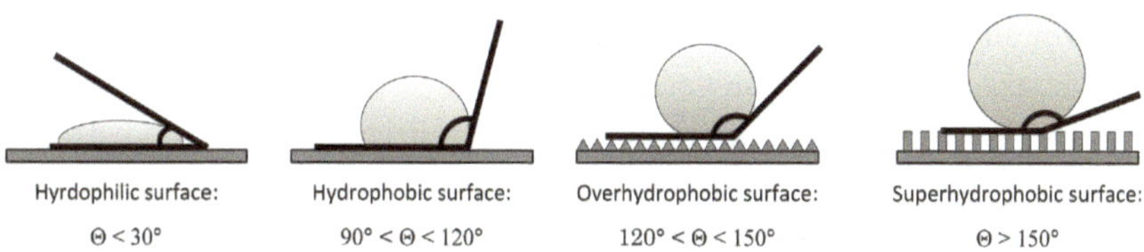

Fig. 3.10 Concept of superhydrophobic hybridization of concrete pore surface [14].

3.10 Biomimetics and engineering superhydrophobic surfaces

Biomimetics, also known as biomimicry and bionics, can be defined as "the abstraction of good design from nature" and came about when engineers and medical researchers realized that "many of the answers they were seeking were already available in nature, refined to near-perfection during millions of years of evolution" [44–46]. The engineering of superhydrophobic and self-cleaning surfaces found on the leaves of the lotus plant (*Nelumbo nucifera*) is an example of how biomimetics can be used for effective engineering design (Fig. 3.10). The superhydrophobicity of most plant surfaces is achieved by hierarchical structures of convex or papilla epidermal cells with three-dimensional wax structures on top [44]. The hierarchical (double structured) surface is a characteristic of the lotus leaf, which is built up of convex cells and a much smaller superimposed layer of hydrophobic three-dimensional wax tubules. Wetting of such surfaces is minimized because air is trapped in the cavities of the convex cell sculptures and the hierarchical roughness enlarges the water-air interface while reducing the solid-water interface. Water on such a surface gains very little energy through adsorption and forms a spherical droplet (Fig. 3.10, middle-right), and the contact area and the adhesion to the surface are dramatically reduced [12,14]. A nature-inspired approach can be used to improve the performance of well-known hydrophobic materials and, therefore, to control the wettability of solid materials. Superhydrophobic surfaces with a water contact angle Θ larger than 150° (Fig. 3.10, right) have generated much interest due to their potential in industrial applications (mainly for self-cleaning application) but have not been realized yet for the enhancement of concrete durability.

In order to realize the superhydrophobic hybridization of concrete inter-pore space, the siloxane-based admixture (e.g., PEHS/PMHS) is combined with small quantities of superfine, submicron-, or nanosized particles such as nanosilica, nanoclay additives, or SiO_2-rich reactive powders [5,12,14]. Fig. 3.11 shows that the modified PEHS/PMHS admixture forms small, uniform air voids that are evenly distributed within the cement paste. As a result, the surface of these voids is covered by the hydrophobic particles, providing the effect of superhydrophobic hybridization. The bond between the particles and cement pore surface depends on the size and reaction ability of particles and also on the degree of particle hydrophobization. It can be envisioned that even loosely attached particles will provide sufficient pillars

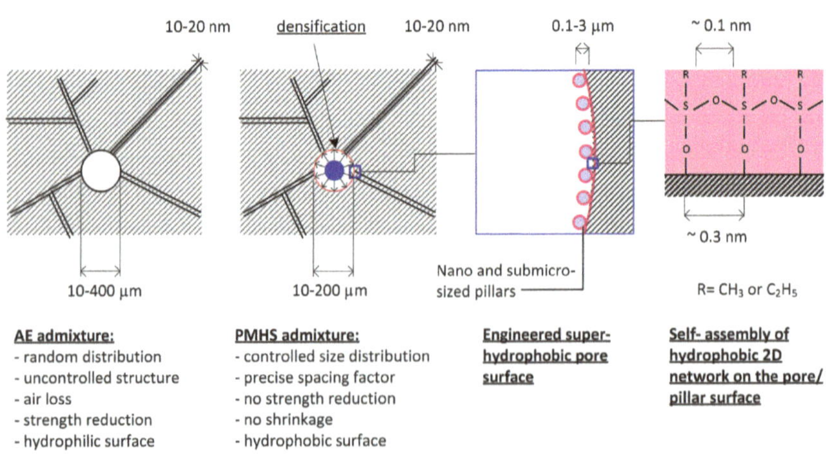

Fig. 3.11 How superhydrophobic hybridization of concrete works [12].

and roughness required for the superhydrophobic effect; this will limit the transport of water or aggressive solutions, thereby boosting concrete resistance to detrimental chemicals.

Hydrophobized polyvinyl alcohol (PVA)-fiber-reinforced mortar specimens based on PC and slag cement were tested for freezing and thawing. Polymethyl-hydrogen siloxane (PHMS) was used as the hydrophobic agent. Microparticles of SF or metakaolin were used to stabilize [12,14,47] and modify the PMHS emulsion using the "shell" and "core" approach [12]. The fiber-reinforced concrete (FRC) was designed with 2.75% (by volume) of PVA fibers, water-to-cementitious material ratio of 0.30 and 0.45 with sand-to-cementitious material ratio of 0.5 and 1.0, respectively. These compositions were compared with hydrophobic composites.

The effects of fibers, the concentration of cementitious material, and the concentration of the hydrophobic agent (emulsion type) were investigated for the contact angle. Samples treated with a higher concentration of PHMS emulsion showed greater contact angle than lower concentration and combining fiber, submicron- and nanosized particles of SF, and PHMS treatment resulted in samples with further increased contact angle due to the micro/nanoroughness of the hierarchical roughness as predicted (Fig. 3.11). Results of the freeze-thaw testing up to 750 cycles at −50°C demonstrate that the main factor affecting the durability of FRC is the W/C (Fig. 3.12).

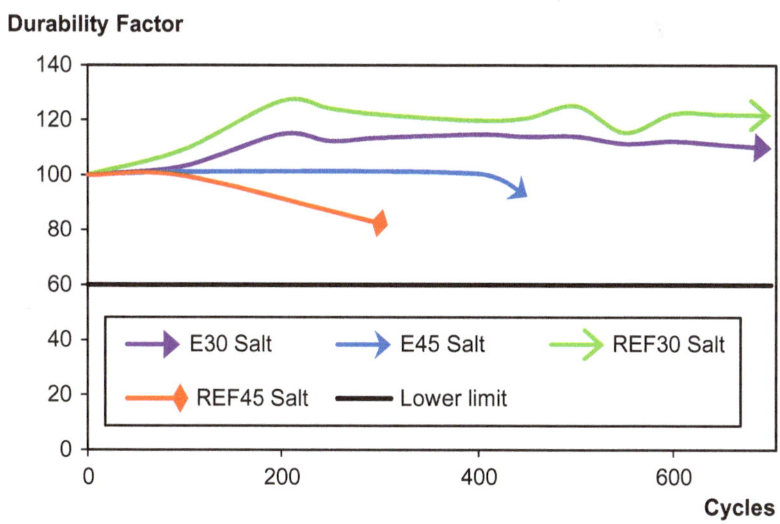

Fig. 3.12 Durability factor for samples during freezing and thawing in 5% NaCl solution [14].

All samples with a W/C of 0.3 displayed excellent performance through as many as 700 cycles by maintaining of durability factor of 100 or more. For these samples, an increase in the durability factor can be seen through the first 200 cycles. This can be attributed to the continued strength gain (since the samples were placed in the chamber at the age of 14 days). This, however, was not the case for FRC with higher W/C, which demonstrated that weaker matrix with larger capillary porosity was subjected to immediate deterioration due to severe exposure (cycling in 5% NaCl solution). The FRC with PMHS emulsions (E45) demonstrated a better performance versus the reference mixture. Here, the FRC samples prepared with a higher W/C and superhydrophobic admixtures (E45 Salt) displayed better performance compared to the samples without (REF45 salt).

The application of hydrophobic emulsions with nanosilica as a coating on mortar tiles proved to result in high contact and low roll-off angles, especially when diluted emulsions were used [47–52]. Based on the durability-testing results, it is clear that FRC with low W/C had an ultimate freeze-thaw response demonstrating a durability factor of more than 100% through as many as 700 freezing and thawing cycles in 5% NaCl at 50°C. The effect of superhydrophobic admixtures is evident in FRC tested in NaCl solutions. In high W/C (~0.45) FRC, the use of superhydrophobic emulsions enables improvement to freeze-thaw resistance. The combination of surface and internal hydrophobization should result in ultradurable cementitious composites [13,14].

3.11 Developing new functionalities

Concrete represents one of the oldest material technologies that have widespread use in civil and infrastructure applications due to its high compressive strength and exceptional durability that can be achieved at relatively low production costs [43]. With the intensification of urbanization, a considerable CO_2 footprint of commercial and residential buildings is associated with ever-increasing levels of energy consumption, thereby requiring innovative approaches to the direct production of electrical energy. Therefore, realizing solar energy harvesting capabilities of concrete surfaces can offer multiple benefits, including onsite power generation. Most of the recent research in concrete was focused on improving the mechanical characteristics; however, new pathways to develop concrete with novel value-added functionalities were suggested [22,50,53–55]. Examples include incorporating

TiO_2 nanoparticles in photocatalytic concrete that is capable of air depollution (e.g., by reducing NO_x upon exposure to UV light) and self-cleaning [50,52].

The feasibility of the electrolyte-based concrete-dye cell (EC-DC) based on a multilayer sandwich structure composed of TiO_2-concrete and the conductive carbon nanotube cementitious composite of a controlled porous structure was evaluated [53]. The EC-DC current-voltage measurements performed under illumination at incident optical powers of ~46 mW confirmed the generation of electrical power of ~0.64 μW with almost half generated via battery effect. The reported research presents a step forward in developing multifunctional, zero CO_2 footprint concrete product derivatives with novel solar-to-electrical power conversion capabilities.

3.12 Conclusions

The various described methods of nanomodification are effective to boost the mechanical performance, durability, and SCM utilization rate that can provide sustainable solutions for the concrete industry. The combination of these methods can result in a flexible, effective, economical, and sustainable solution for producing eco-cement and concrete with high volumes of byproducts such as granulated blast furnace slag, fly ash, or even off-spec materials. The advantage of this proposed solution is its applicability to cement and concrete technology and also its ability to control the properties of nanocement. The described concepts can result in optimal solution to meet the development objectives for sustainable cement materials required in future construction projects.

Nanotechnology has changed and will continue to change our vision, expectations, and abilities to control the material world. These developments will greatly affect construction and the field of construction materials. The major achievements in this domain include:

- the ability to observe the structure at its atomic level and measure the strength and hardness of micro- and nanoscopic phases of composite materials;
- discovery of a highly ordered crystal nanostructure of "amorphous" C—S—H gel;
- self-cleaning concrete based on photocatalyst technology;
- development of finishing materials with self-cleaning properties, discoloration resistance, anti-graffiti protection, and high scratch-and-wear resistance;

- nanometer-thin coatings that protect carbon steel against corrosion and enhance thermal insulation of window glass; and
- smart stress-sensing composites.

Among new nano-engineered polymers are highly efficient superplasticizers for concrete and high-strength fiber composites with exceptional energy-absorbing capacity. Nanoparticles, such as silicon dioxide, were found to be a very effective additive to polymers and concrete, a development recently realized in high-performance and self-compacting concrete with improved workability and strength.

Concrete, one of the largest commodities consumed by humankind, has significant, but not completely explored, potential. Better understanding and precise engineering of an extremely complex structure of cement-based materials at the nanolevel will result in a new generation of concrete that is stronger and more durable, with desired stress-strain behavior and possibly possessing the range of newly introduced "smart" properties, such as electrical conductivity, temperature-, moisture-, and stress-sensing abilities. At the same time, this new concrete should be sustainable, cost, and energy effective—in essence exhibiting the qualities modern society demands. Nanobinders or nano-engineered cement-based materials with nanosized cementitious component or other nanosized particles are the next ground-breaking development.

The development of superhydrophobic and nanobinder-based concrete can set the top of sustainability benchmarks and so can serve as the next technological platform for sustainable development of the concrete industry by reducing the use of cement content by at least 50% and radically extending the service life of developed concrete by a factor of 5 and higher. Such technology represents a multi-scale design involving superhydrophobic hybridization and the fine-tuning of the nano- and microstructure of cement-based materials to meet these performance specifications. The proposed concept serves as a platform for the development of new types of concrete, including those with ultra-high durability and increased volumes of utilized waste and industrial byproducts. In addition, concrete with higher volumes of byproducts can provide greener solutions for the next generation of eco-concrete at lower costs. Reducing PC content by 50% and improving the overall performance of eco-concrete can help to reduce the carbon footprint of the structures by 40% and decrease the use of raw materials required for cement and concrete manufacture [1,3,4,6,15,17,18,20,31].

Additionally, the introduction of these novel materials into the public sphere through civil infrastructure will necessitate an evaluation and understanding the impact they may have on the environment and human health.

Nanotechnology is still in its pre-exploration stage—it is just emerging from fundamental research onto the industrial practice; thus, its full-scale applications—especially in construction in 2022—are still limited. However, the tremendous potential for nanotechnology to improve the performance of conventional materials and processes is most promising.

References

[1] K. Sobolev, Sustainable development of the cement industry and blended cements to meet ecological challenges, Sci. World J. 3 (2003) 308–318, https://doi.org/10.1100/tsw.2003.23.

[2] Cement, United States Geological Survey, National Minerals Information Center, 2019.

[3] M.A. Robbie, Global CO_2 emissions from cement production, Earth Syst. Sci. Data 10 (2018) 195–217, https://doi.org/10.5194/essd-10-195-2018.

[4] M.L. Marceau, M.A. Nisbet, M.G. VanGeem, Life Cycle Inventory of Portland Cement Manufacture, PCA R&D Serial No. 2095b, 2006, p. 68.

[5] K. Sobolev, V. Batrakov, The effect of a PEHSO on the durability of concrete with supplementary cementitious materials, ASCE J. Mater. Civil Eng. 19 (10) (2007) 809–819.

[6] M. Corradi, R. Khurana, R. Magarotto, Controlling performance in ready mixed concrete, Concr. Int. 26 (8) (2004) 123–126.

[7] K. Sobolev, A. Yeginobali, The development of high-strength mortars with improved thermal and acid resistance, Cem. Concr. Res. 35 (3) (2005) 578–583.

[8] S.W. Forster, High-performance concrete - stretching the paradigm, Concr. Int. 16 (10) (1994) 33–34.

[9] K. Sobolev, S. Muzenski, I. Flores-Vivian, Ultra-high strength cement-based composites designed with aluminum oxide nano-fibers, Constr. Build. Mater. 220 (2019) 177–186.

[10] K. Sobolev, The development of a new method for the proportioning of high-performance concrete mixtures, Cem. Concr. Compos. 26 (7) (2004) 901–907.

[11] I. Jansson, U. Skarp, C. Bigley, The value of colloidal silica for enhanced durability in high fluidity cement based mixes, in: 5th International RILEM Symposium on Self-Compacting Concrete, 2007.

[12] I. Flores-Vivian, V. Hejazi, M. Kozhukhova, M. Nosonovsky, K. Sobolev, Self-assembling particle-siloxane coatings for superhydrophobic concrete, ACS Appl. Mater. Interfaces 5 (24) (2013) 13284–13294.

[13] S. Muzenski, I. Flores-Vivian, K. Sobolev, Hydrophobic modification of ultra-high-performance fiber-reinforced composites with matrices enhanced by aluminum oxide nano-fibers, Constr. Build. Mater. 244 (2020), 118354.

[14] S. Muzenski, I. Flores-Vivian, K. Sobolev, Hydrophobic engineered cementitious composites for highway applications, Cem. Concr. Compos. 57 (2015) 68–74.

[15] P. Serna, F.L. Monte, E.J. Mezquida-Alcaraz, E. Cuenca, V. Mechtcherine, M. Reichardt, A. Peled, O. Regev, R.P. Borg, A. Tretjakov, D. Lizunov, K. Sobolev, S. Sideri, K. Nelson, E.M.G. Brac, L. Ferrara, Upgrading the concept of

UHPFRC for high durability in the cracked state: the concept of ultra high durability concrete (UHDC) in the approach of the H2020 project RESHEALIENCE, in: Sustainable Materials Systems and Structures - SMSS 2019, vol. 128, RILEM Publications SARL, 2019, pp. 764–771.
[16] M. Moini, I. Flores-Vivian, A. Amirjanov, K. Sobolev, The optimization of aggregate blends for sustainable low cement concrete, Constr. Build. Mater. 93 (2015) 627–634.
[17] K. Sobolev, P. Türker, S. Soboleva, G. Iscioglu, Utilization of waste glass in eco-cement: properties and microstructural observations, Waste Manag. 27 (7) (2007) 971–976.
[18] K. Sobolev, M. Arikan, High volume mineral additive ECO-cement, Am. Ceram. Soc. Bull. 81 (1) (2002) 39–43.
[19] B. Langan, K. Weng, M. Ward, Effect of silica fume and fly ash on heat of hydration of Portland cement, Cem. Concr. Res. 32 (2002) 1045–1051.
[20] K. Sobolev, T.R. Naik, Performance as a factor for sustainability of the cement industry, in: CANMET/ACI International Symposium on Sustainable Development of Cement and Concrete, Toronto, Canada, 2005, pp. 295–312.
[21] K. Sobolev, M. Ferrada-Gutiérrez, How nanotechnology can change the concrete world: part 1, Am. Ceram. Soc. Bull. 10 (2005) 14–17.
[22] K. Sobolev, M. Ferrada-Gutiérrez, How nanotechnology can change the concrete world: part 2, Am. Ceram. Soc. Bull. 11 (2005) 16–19.
[23] C. Plassard, E. Lesniewska, I. Pochard, A.A. Nonat, Investigation of the surface structure and elastic properties of calcium silicate hydrates at the nanoscaled, Ultramicroscopy 100 (3–4) (2004) 331–338.
[24] A. Gmira, J. Minet, A. Francischini, N. Lequeux, R.J.-M. Pellenq, H. Van Damme, Molecular engineering of the cohesion in neat and hybrid cement hydrates, in: K. Sobolev, S.P. Shah (Eds.), Proceedings of the ACI Session on Nanotechnology of Concrete: Recent Developments and Future Perspectives – 2006, Denver, USA, ACI SP-254, 2008, pp. 29–40.
[25] K. Sobolev, S.P. Shah (Eds.), Proceedings of the ACI Session on Nanotechnology of Concrete: Recent Developments and Future Perspectives - 2006, Denver, USA, ACI SP-254, 2008.
[26] M. Collepardi, J. Ogoumah, U. Skarp, R. Troli, Influence of amorphous colloidal silica on the properties of self-compacting concretes, in: Proceedings of the International Conference, Challenges in Concrete Construction - Innovations and Developments in Concrete Materials and Construction, Dundee, UK, 2002.
[27] F. Sanchez, K. Sobolev, Nanotechnology in concrete – a review, Constr. Build. Mater. 24 (11) (2010) 2060–2071.
[28] I. Flores, K. Sobolev, L.M. Torres-Martinez, E.L. Cuellar, P.L. Valdez, E. Zarazua, Performance of cement systems with Nano-SiO_2 particles produced by using the sol–gel method, Transp. Res. Rec. 2141 (1) (Jan. 2010) 10–14, https://doi.org/10.3141/2141-03.
[29] I. Flores-Vivian, R.G.K. Pradoto, M. Moini, M. Kozhukhova, V. Potapov, K. Sobolev, The effect of SiO_2 nanoparticles derived from hydrothermal solutions on the performance of Portland cement based materials, Front. Struct. Civ. Eng. 11 (4) (2017) 436–445, https://doi.org/10.1007/s11709-017-0438-2.
[30] K. Sobolev, et al., Nano-engineered cements with enhanced mechanical performance, J. Am. Ceram. Soc. 99 (2015) 564–572.
[31] M. Collepardi, S. Collepardi, U. Skarp, R. Troli, Optimization of silica fume, fly ash and amorphous nano-silica in superplasticized high-performance concretes, in: Proceedings of 8th CANMET/ACI International Conference on Fly Ash, Silica Fume, Slag and Natural Pozzolans in Concrete, SP-221, Las Vegas, USA, 2004, pp. 495–506.

[32] J.S. Belkowitz, D. Armentrout, An Investigation of Nano Silica in the Cement Hydration Process, NRMCA, 2017.
[33] J. Björnström, A. Martinelli, A. Matic, L. Börjesson, I. Panas, Accelerating effects of colloidal nano-silica for beneficial calcium–silicate–hydrate formation in cement, Chem. Phys. Lett. 392 (1–3) (2004) 242–248.
[34] Q. Ye, Z. Zhang, D. Kong, R. Chen, Influence of nanosilica addition on properties of hardened cement paste as compared with silica fume, Constr. Build. Mater. 21 (3) (2007) 539–545.
[35] A. Porro, J.S. Dolado, I. Campillo, E. Erkizia, Y. de Miguel, Y.S. de Ibarra, A. Ayuela, Effects of nanosilica additions on cement pastes, in: Applications of Nanotechnology in Concrete Design, 2005, pp. 87–96.
[36] J.J. Gaitero, I. Campillo, A. Guerrero, Reduction of the calcium leaching rate of cement paste by addition of silica nanoparticles, Cem. Concr. Res. 38 (8–9) (2008) 1112–1118.
[37] B. Green, Development of a high-density cementitious rock-matching grout using nano-particles, in: Proceedings of ACI Session on "Nanotechnology of Concrete: Recent Developments and Future Perspectives, SP-254", 2008.
[38] J.J. Thomas, et al., Influence of nucleation seeding on the hydration mechanisms of tricalcium silicate and cement, J. Phys. Chem. C 113 (2009) 4327–4334.
[39] K. Sobolev, Mechano-chemical modification of cement with high volumes of blast furnace slag, Cem. Concr. Compos. 27 (7–8) (2005) 848–853.
[40] K. Sobolev, Z. Lin, Y. Cao, H. Sun, I. Flores-Vivian, T. Rushing, T. Cummins, W.J. Weiss, The influence of mechanical activation by vibro-milling on the early-age hydration and strength development of cement, Cem. Concr. Compos. 71 (2016) 53–62.
[41] S. Sinyoung, K. Kunchariyakun, S. Asavapisit, K.J.D. MacKenzie, Synthesis of belite cement from nano-silica extracted from two rice husk ashes, J. Environ. Manag. 190 (Apr. 2017) 53–60, https://doi.org/10.1016/j.jenvman.2016.12.016.
[42] H.H. Bache, Densified cement/ultrafine particle based materials, in: 2nd International Conference on Superplasticizers in Concrete, Ottawa, 1981.
[43] J.J. Biernacki, J.W. Bullard, G. Sant, K. Brown, F.P. Glasser, S. Jones, T. Ley, R. Livingston, L. Nicoleau, J. Olek, F. Sanchez, R. Shahsavari, P.E. Stutzman, K. Sobolev, T.P. Prater, Cements in the 21st century: challenges, perspectives, and opportunities, J. Am. Ceram. Soc. 100 (7) (2017) 1551–2916.
[44] Y. Liu, Superhydrophobic surfaces from a simple coating method: a bionic nanoengineering approach, Nanotechnology 17 (2006) 3259–3263.
[45] M. Nosonovsky, Multiscale roughness and stability of superhydrophobic biomimetic interfaces, Langmuir 23 (6) (2007) 3157–3161.
[46] K. Koch, B. Bhushan, W. Barthlott, Diversity of structure, morphology and wetting of plant surfaces, Soft Matter 4 (2008) 1943–1963.
[47] V. Hejazi, K. Sobolev, M. Nosonovsky, From superhydrophobicity to icephobicity: forces and interaction analysis, Nat. Sci. Rep. 3 (2013) 2194.
[48] G. Moriconi, F. Tittareli, Effectiveness of surface or bulk hydrophobic treatments in cementitious materials, in: Protection of Historical Buildings: Prohitech 09, 2009, pp. 1071–1075.
[49] M. Raupach, L. Wolff, Investigations on long-term durability of hydrophobic treatment on concrete, Surf. Coat. Int. B: Coat. Trans. 99 (2) (2005) 127–133.
[50] S. Lanka, E. Alexandrova, M. Kozhukhova, M.S. Hasan, M. Nosonovsky, K. Sobolev, Tribological and wetting properties of TiO_2 based hydrophobic coatings for ceramics, J. Tribol. 141 (10) (2019).
[51] R. Ramachandran, M. Kozhukhova, K. Sobolev, M. Nosonovsky, Anti-icing superhydrophobic surfaces: controlling entropic molecular interactions to design novel icephobic concrete, Entropy 18 (4) (2016) 132.

[52] E.E. Hekal, M. Abd-El-Khalek, Mechanical and physico-chemical properties of hardened Portland cement pastes containing hydrophobic admixtures part 1: compressive strength and hydration kinetics, ZKG Int. 52 (12) (1999) 697–700.
[53] J. Hoheneder, I. Flores-Vivian, Z. Lin, P. Zilberman, K. Sobolev, The performance of stress-sensing smart fiber reinforced composites in moist and sodium chloride environments, Compos. Part B 73 (2015) 89–95.
[54] M. Faraldos, R. Kropp, M.A. Anderson, K. Sobolev, Photocatalytic hydrophobic concrete coatings to combat air pollution, Catal. Today 259 (2016) 228–236.
[55] T. Hosseini, I. Flores-Vivian, K. Sobolev, N. Kouklin, Concrete embedded dye-synthesized photovoltaic solar cell, Nat. Sci. Rep. 3 (2013) 2727, https://doi.org/10.1038/srep02727.

CHAPTER 4

Design, mechanism, and performance of cement-based materials with zero-dimensional nanomaterials

Zemei Wu[a,b] and Kamal H. Khayat[b]
[a]College of Civil Engineering, Hunan University, Changsha, People's Republic of China
[b]Center for Infrastructure Engineering Studies, Department of Civil, Architectural and Environmental Engineering, Missouri University of Science and Technology, Rolla, MO, United States

4.1 Introduction

Concrete is an intrinsically heterogeneous material consisting of cementations materials, water, fine aggregates, coarse aggregates, chemical admixtures, and/or fibers. Its macro-properties, including mechanical properties, dimensional stability, and durability, are dominated by the structure at the micro and nanoscale levels. Pores, interface, and atomic irregularities of various types, existing at micro and nanoscale levels can degrade the performance of concrete [1,2]. For example, the quality of the interfacial transition zone (ITZ) between cementitious matrix and aggregate and/or fiber exerts a great effect on permeability and mechanical properties. This is because the interface is more porous and less stiff than that of the bulk cementations matrix. The existence of pores and microcracks in this zone with steel reinforcements and aggregate will trigger concrete failure because less ultimate stress is required to initiate the cracking. Pore structure among gel pores also significantly affects autogenous shrinkage of concrete, thus affecting the cracking and durability of concrete.

Portland cement clinker contains four principal minerals, namely, alite (C_3S), belite (C_2S), aluminate (C_3A), and ferrite (C_4AF). The underlying core to bind all components of cement-based materials to finally lead to paste, mortar, and concrete is mainly attributed to complex chemical reactions between cement and water. When cement grains are contacted with water, various chemical reactions occur simultaneously. As the reactions proceed, the products of the hydration process gradually bond together sand, gravel particles, and other components of cement-based materials to form a

https://doi.org/10.1016/B978-0-12-817832-4.00009-2

rigid system. The resulting porous multiphase matrix contains the main reaction products C—S—H gel, calcium hydroxide, AFm and AFt phases, and unhydrated clinker. The main hydration product of calcium silicate hydrate phase (C—S—H gel), occupying at least 50% to 60% of the concrete volume, is the most important substance dominating the properties of cement-based materials. It is reported as a nano-structured material with an average diameter of approximately 10 nm [3]. The quantity and quality of C—S—H can have important implications for mixture design and overall performance of cement-based materials. C—S—H can be classified into at least three structurally different forms based on its intrinsic properties, i.e., stiffness, hardness, and volume fractions, including low-density, high-density, and ultra-high-density C—S—H [4–6].

Use of low water-to-binder ratio, high range water reducer, fine particles, such as supplementary cementitious materials (SCMs) and nanoparticles, and decreased size of aggregates, have been employed to develop high-performance concrete (HPC) and ultra-high-performance concrete (UHPC). Among these various measures being explored, the use of nanoparticles is one of the most promising strategies given their strengthening effect from the nanoscale. Therefore, providing a scientific basis through exploring nanoscience and nanotechnology will lead to sustainable and eco-efficient construction. Also, this approach will link the initial synthesis and processing conditions with the evolving microstructure and the resulting macroscopic response of cement-based materials.

Since the introduction of nanomaterials, extensive research has been conducted to promote their use in cement-based materials. It is well known that nanomaterials can act as nuclei for cement phases, further promoting cement hydration due to their high reactivity. Its filler effect can fill pores among coarse particles, thus leading to densification of microstructure, enhancement in homogeneity, and improvement in strength and durability.

This chapter will discuss the common zero-dimensional (0-D) nanomaterials employed in cement-based materials. The dispersion and mixing procedure, mechanisms underlying performance enhancement, and how those materials affect the fresh and hardened properties of cement-based materials will be explored.

4.2 Types and characteristics of 0-D nanomaterials

Nanomaterials are engineered materials processed to have an extremely small size, typically from 1 nm to 100 nm, to take advantage of their unique physical and chemical properties. The technologies that lead to the industrial outputs of nanomaterials involve plasma arching, chemical vapor deposition, electro-deposition, sol-gel synthesis, mechanical attrition, and the use of natural nanosystems [7]. Based on their geometries, nanomaterials can be categorized into 0-D, one-dimensional (1-D), two-dimensional (2-D), and three-dimensional nanomaterials (3-D), as depicted in Fig. 4.1. 0-D nanomaterials include nanocluster materials and nanodispersions, i.e., materials in which nanoparticles are isolated from each other. Fig. 4.2 compares the particle size and specific surface area (SSA) of typically nanomaterials to commonly used components in concrete. The 0-D nanomaterials are one to three orders of magnitude finer than those of cement and SCMs, while four to six orders of magnitude finer compared to the size of aggregates. The use of different scales of materials combined with chemical admixtures, such as high-range water-reducer, results in various types of concrete.

The 0-D nanomaterials typically used in cementitious materials mainly include nano-SiO_2, nano-$CaCO_3$, nano-Fe_2O_3, nano-Fe_3O_4, nano-TiO_2, nano-CuO, and nano-ZiO_2. Table 4.1 summarizes the physical

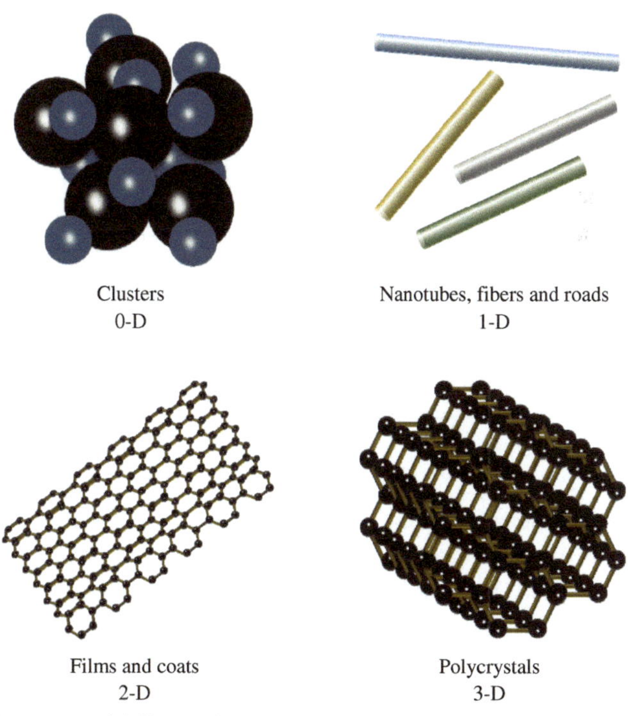

Fig. 4.1 Illustration of different dimensional types of nanomaterials.

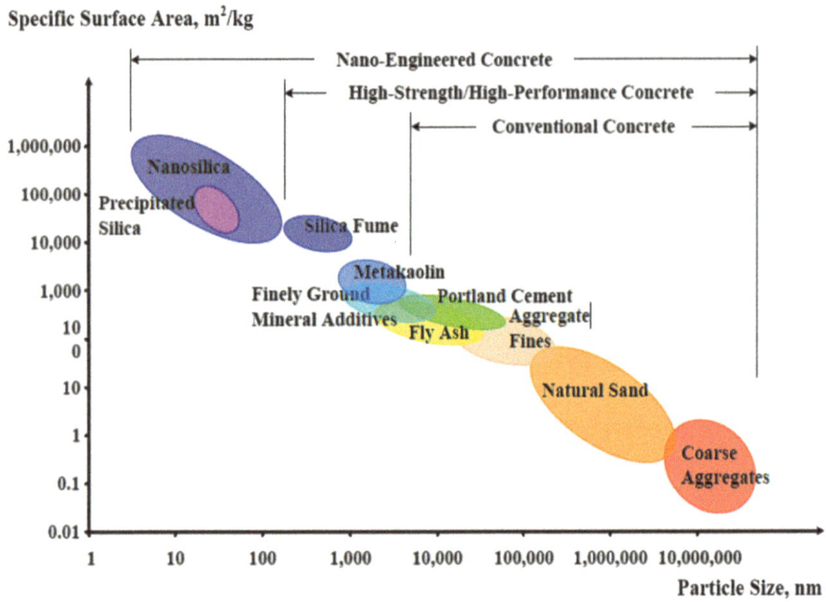

Fig. 4.2 Particle size and SSA related to concrete materials [8].

Table 4.1 Physical and chemical composition of typical 0-D nanomaterials.

Type	Density (g/cm^3)	Diameter (nm)	SSA (m^2/g)	Color	Morphology
Nano-SiO_2	2.2–2.6	10–20	50–200	White	Amorphous
Nano-$CaCO_3$	2.9	10–100	30–60	White	Cubic or hexagonal
Nano-Fe_2O_3	5.2	20–40	40–80	Red or reddish-brown	Spherical
Nano-Fe_3O_4	4.8–5.1	15–30	40–160	Dark brown	Spherical
Nano-TiO_2	4.2	10–500	8–60	White	Spherical
Nano-Al_2O_3	4.0	10–300	5–15	White	Rhombohedra

and chemical properties of these 0-D nanomaterials. The density, color, and morphology can vary from each other, depending on the material type and production method. Most of the 0-D nanomaterials exhibit spherical particle shape, as reflected from scanning electron microscope (SEM) observation in Table 4.2.

However, some 0-D nanomaterials show their unique shapes. For instance, nano-$CaCO_3$ is a cubic or hexagonal substance, while

Table 4.2 SEM images and XRD analysis of commonly used 0-D nanoparticles.

Type	SEM image	XRD analysis	Reference
Nano-SiO_2	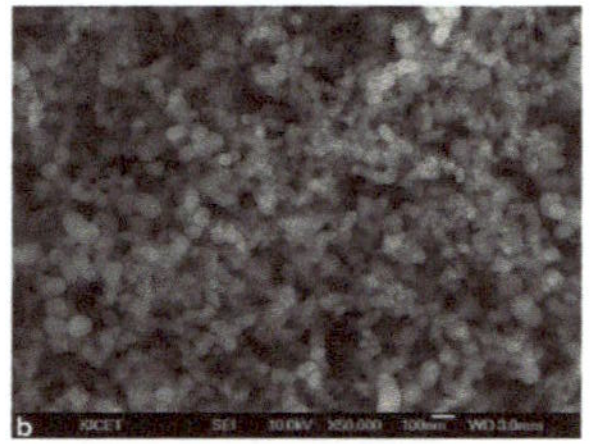	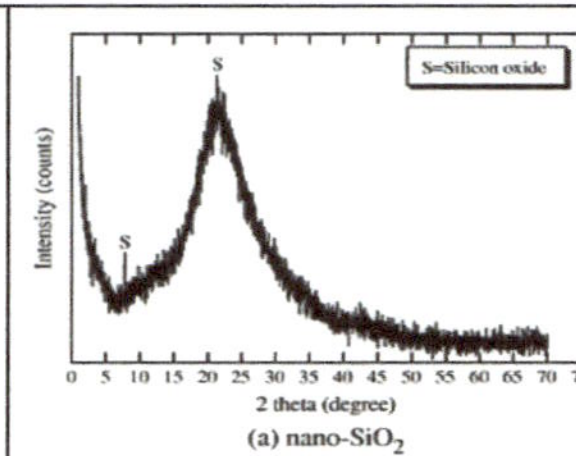	[9,10]
Nano-$CaCO_3$		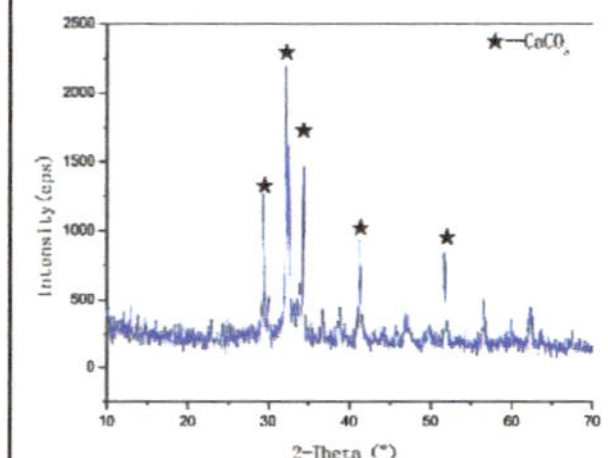	[11]
Nano-Fe_2O_3	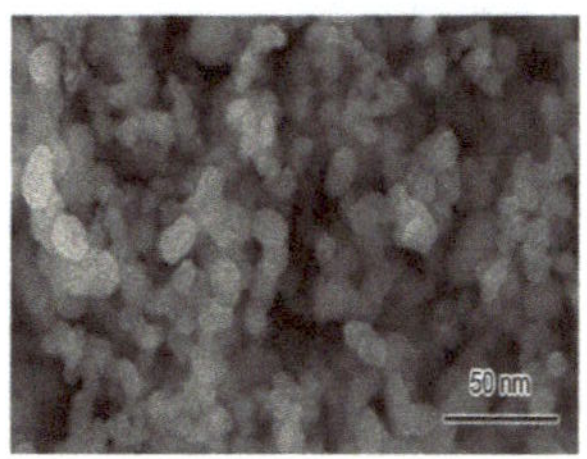	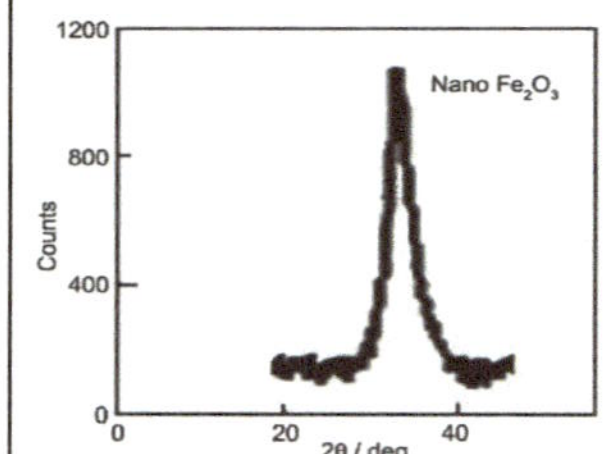	[12]

Continued

Table 4.2 SEM images and XRD analysis of commonly used 0-D nanoparticles—cont'd

Type	SEM image	XRD analysis	Reference
Nano-Fe_3O_4		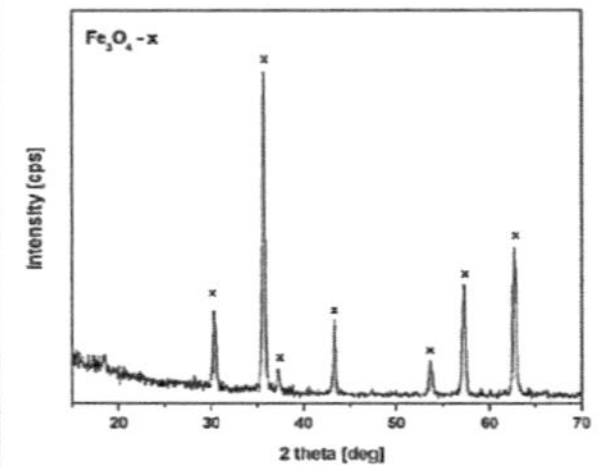	[13]
Nano-TiO_2	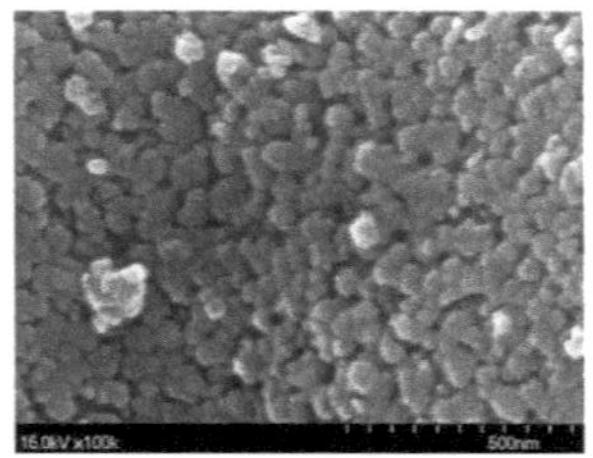	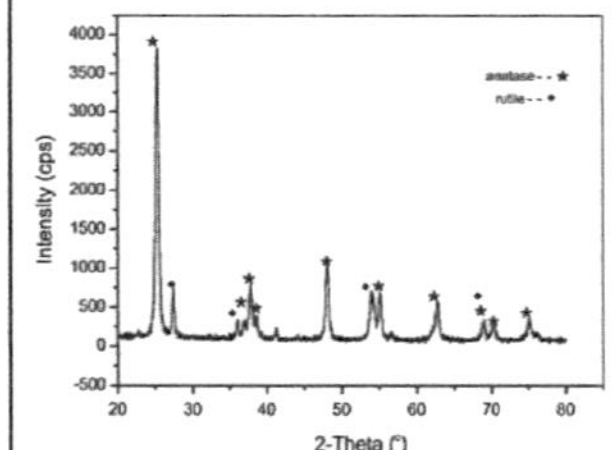	[14]
Nano-Al_2O_3		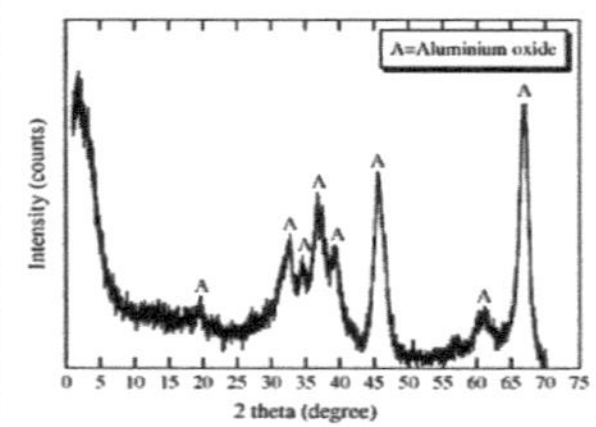(b) nano-Al_2O_3	[10,15]

nano-Al_2O_3 is a rhombohedron one. The compounds of the nanomaterials can be identified using X-ray diffraction (XRD) analysis. The intensities of the diffracted waves depend on the type and arrangement of atoms in the crystal structure.

4.3 Dispersing and mixing of 0-D nanomaterials

Effective dispersion of nanoparticles in cement-based materials is quite challenging because of the great surface energy associated with their high fineness and SSA, which can easily lead to agglomeration issues. The agglomerated nanoparticles can overlap with each other to not be available for chemical reaction, leading to decreased utilization efficiency. Meantime, it can lead to weak zones inside cement-based materials, such as entrapped air voids, which are recognized as trigger zones for concrete failure when subjected to stress concentration. Therefore, the dispersion state of nanoparticles can significantly affect the key fresh and hardened properties of cement-based materials. To maximize their function, nanomaterials should be dispersed appropriately before mixing with constituents, including cementitious materials, aggregate, chemical admixtures, and water.

In some literature, the 0-D nanoparticles are directly mixed into cementitious materials following the normal mixing procedure with aggregate and then liquids. This is simple and easy to conduct. Satisfactory dispersion can be achieved to contribute to the mechanical properties of cement-based materials at relatively low content [16]. However, agglomeration issues are unavoidable if the 0-D nanoparticles are added at relatively high content. Attempts, such as physical and chemical dispersion methods, have been employed to secure a more uniform dispersion of 0-D nanoparticles. The 0-D nanoparticles can be predispersed in water or a combination of water and chemical dispersing agents using ultra-sonification [17,18]. For example, nanoparticles can be stirred in water at a rotation speed of 120 rpm for approximately several to 30 min before adding polycarboxylate superplasticizer [19,20]. Water-reducing agents, such as b-naphthalene sulfonic acid and formaldehyde condensates, are also used to help disperse the 0-D nanoparticles in cement paste and achieve good workability [21]. In this case, the mixing of materials would follow the following procedure. Firstly, water and/or water-reducing agent are first mixed using a mortar mixer. The 0-D nanoparticles are then added and stirred at high speed ($\sim$ 150 rpm) for several minutes. Cementation materials, sand, and coarse aggregate are then mixed at a low speed for several minutes in a centrifugal concrete blender.

Finally, the mixture of water, water-reducing agent, and 0-D nanoparticle are slowly added and stirred at a relatively low speed for several minutes to achieve good workability. The stirring duration depends on the dispersion state of the materials.

Fig. 4.3 shows typical SEM and transmission electron microscopy (TEM) images of nano-TiO_2 particles before and after treatment using 28 kHz ultrasonic waves in water for 10 min. The nano-TiO_2 particles are fully overlapped with each other before dispersion (Fig. 4.3). The flocculation of nano-TiO_2 particles is effectively mitigated after ultra-sonification treatment (Fig. 4.3A and B). Chemical dispersing agents, such as superplasticizer, can facilitate the nanoparticle dispersion in liquid solution due to electrostatic repulsion and steric hindrance [22]. Sonification can be employed

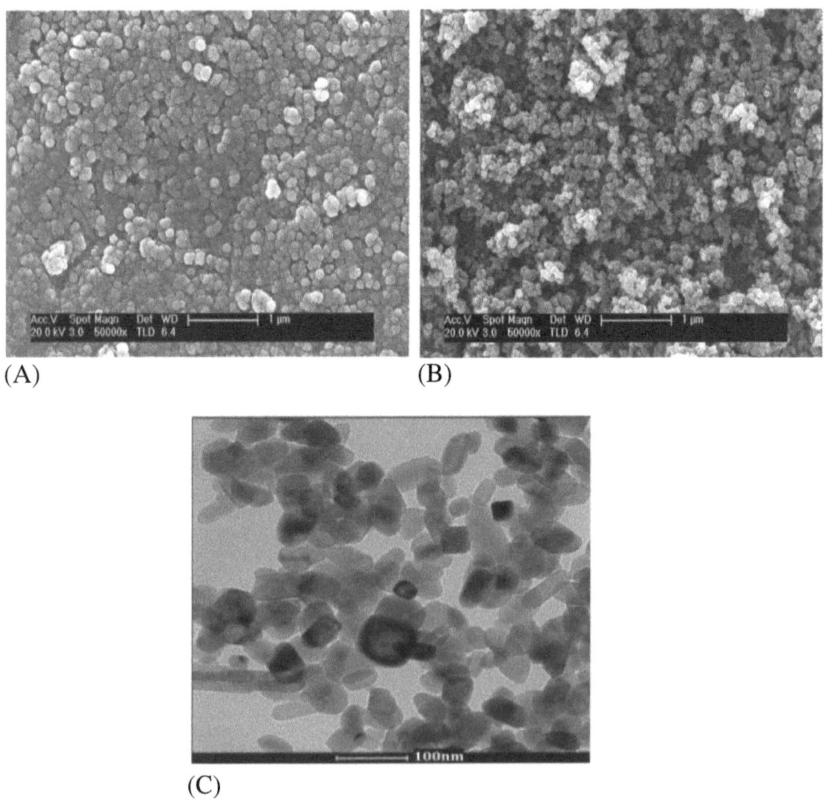

Fig. 4.3 Images of nano-TiO_2 particles before and after dispersion. (A) Nano-TiO_2 particles before dispersion. (B) Nano-TiO_2 particles after dispersion. (C) TEM image after dispersion [20].

to reduce the agglomerate size by approximately an order of magnitude. It should be noted that the use of extensive sonification and dispersing agents can better disperse 0-D nanoparticles. Still, it cannot guarantee well-uniformed dispersion of 0-D nanoparticles due to the ultra-fine size.

4.4 Mechanisms underlying performance enhancement of cement-based materials with 0-D nanomaterials

Because of the extreme fineness and high-specific surface of 0-D nanoparticles, three mechanisms are often referred to enhance the properties of cement-based materials.

4.4.1 Filler effect

Nanomaterials are approximately one to three orders of magnitude finer than that of ordinary Portland cement and traditional SCMs, as shown in Fig. 4.2 [23]. Therefore, nanomaterials can fill the space between coarse and fine grains to render a dense packing of the system. This can reduce the porosity and enhances the densification of concrete. It should be noted that the addition of high 0-D nanoparticle content allows more water absorbed into nanoparticles, which can significantly increase water and chemical admixture demand [24]. It can also aggravate the agglomeration issues that occur around the aggregate and/or fibers because of its high surface area, thus leading to entrapped air voids and porous ITZ [24]. The resulting porous fiber-matrix interface can greatly reduce fiber-matrix bond strength and eventually mechanical properties of UHPC [24].

4.4.2 Nucleation effect

Nucleation is the process where nuclei or seeds act as templates for crystal growth. Homogeneous nucleation occurs when nuclei form uniformly throughout the parent phase, whereas heterogeneous nucleation forms at structural inhomogeneities, such as container surfaces, impurities, grain boundaries, and dislocations [25]. When 0-D nanoparticles are dispersed in fresh cement-based materials, heterogeneous nucleation will initially occur at cement grain boundaries. This can facilitate the hydration of C_3S and the pozzolanic reaction of SCMs due to the thermodynamic drive that comes from the high surface energy. The hydration acceleration effect on cement associated with 0-D nanoparticles is recognized as the primary one to influence the properties of cement-based materials.

The reaction of cement with water is a dissolution-precipitation process. Immediately after the first contact of cement with water, easily soluble components, like alkalis, C_3S, and free lime are dissolved by the surrounding water [26]. This forms the saturated/oversaturated ion solution, which mainly includes ions, such as Ca^{2+}, SiO_4^{4-}, and SO_4^{2-}. Hydration products, such as calcium hydroxide and C—S—H, precipitate on the surface of dissolving C_3S once reaching their saturated points, as illustrated in Fig. 4.4A. The microstructure of cement-based materials at an early age is often relatively loose since limited formed hydration products. With the addition of 0-D nanomaterials, they can act as hydration seeds around formed C—S—H and/or 0-D nanoparticles to form more C-S-H gels during the precipitation process (Fig. 4.4B). The function of using 0-D nanoparticles for nucleation seeds to form C—S—H is found to be almost comparable to that of using C—S—H seeds (Fig. 4.4B). Therefore, it contributes to enhancing the accelerated dissolution and/or hydration of cementitious materials and shortens the dormant period of cement hydration.

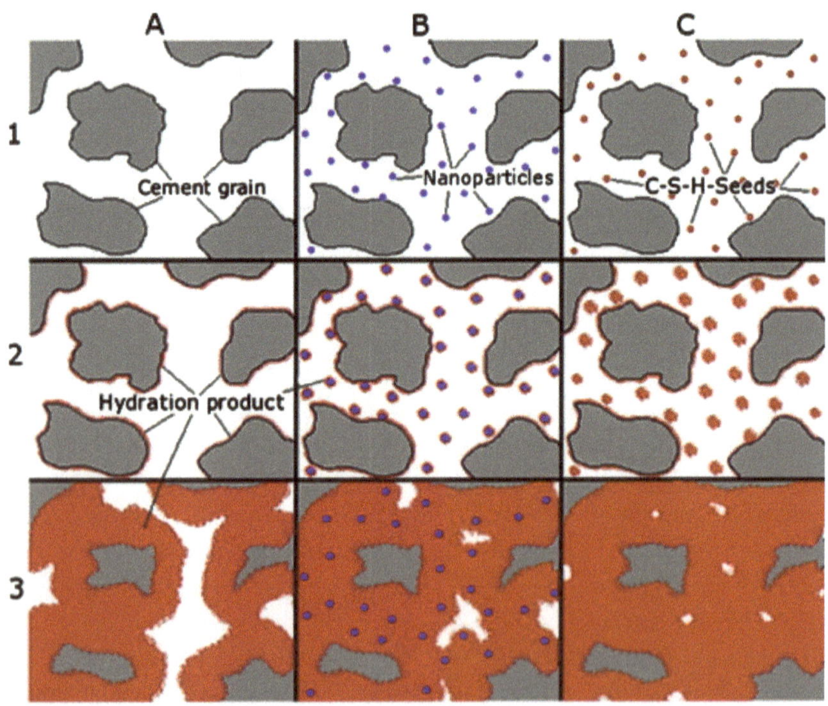

Fig. 4.4 Illustration of hydration process of pure cement (A), with 0-D nanoparticles (B), with C—S—H seeds (C) at different times (1–3) after mixing [27].

4.4.3 Chemical effect

Except for the filler and nucleation effect, some 0-D nanomaterials exhibit chemical effect to form new substances to increase the volume of hydration products. Such chemical reaction can happen between the 0-D nanoparticles and cement clinker and/or hydration products, such as $Ca(OH)_2$ [28,29]. Typical examples are nano-SiO_2 and nano-$CaCO_3$. Such reactivity related to nano-SiO_2 is also called pozzolanic effect, similar to silica fume, due to its rich amorphous SiO_2 content. The pozzolanic reaction is a chemical reaction that involves the consumption of $Ca(OH)_2$ to form C—S—H in the presence of water. $Ca(OH)_2$ crystals in cement-based materials are weak phases because their c-axis are preferentially perpendicular oriented to the surface of rigid phases, such as aggregate and fibers, resulting in entrapment of pores.

When nano-SiO_2 dissolves in water, the unsaturated ≡Si—O— bond would combine with H-OH and/or OH- bonds from water to form a saturated ≡Si—OH bond (Eqs. 4.1, 4.2). The ≡Si—OH bond then reacts with $Ca(OH)_2$ to form C—S—H to decrease the $Ca(OH)_2$ content and increase the C—S—H content (Eq. 4.3).

$$\equiv \mathrm{Si{-}O{-}} + \mathrm{H{-}OH} \rightarrow \equiv \mathrm{Si{-}OH} \tag{4.1}$$

$$\equiv \mathrm{Si{-}} + \mathrm{OH^-} \rightarrow \equiv \mathrm{Si{-}OH} \tag{4.2}$$

$$\equiv \mathrm{Si{-}OH} + \mathrm{Ca(OH)_2} \rightarrow \mathrm{C{-}S{-}H} \tag{4.3}$$

For nano-$CaCO_3$, it can react with C_3A and/or $Ca(OH)_2$ to generate mono- and/or hemicarboaluminate (Eqs. 4.4, 4.5). These can increase the volume of hydrates and eventually result in denser microstructure, greater fiber-matrix bond, and higher mechanical properties of cement-based materials.

$$\mathrm{CaCO_3 + 3CaO \cdot Al_2O_3 + 11\,H_2O \rightarrow 4\,CaO \cdot Al_2O_3 \cdot CaCO_3 \cdot 11H_2O} \tag{4.4}$$

$$\begin{aligned} &\mathrm{0.5CaCO_3 + 11.5H_2O + 3CaO \cdot Al_2O_3 + 0.5Ca(OH)_2} \\ &\quad \mathrm{\rightarrow 4\,CaO \cdot Al_2O_3 \cdot 0.5CaCO_3 \cdot 12H_2O} \end{aligned} \tag{4.5}$$

4.5 Performance of 0-D nanomaterials enabled cement-based materials

4.5.1 Fresh properties

Fresh properties of cement-based materials are essential to evaluate the ease of transport and placement. This can be performed using standard

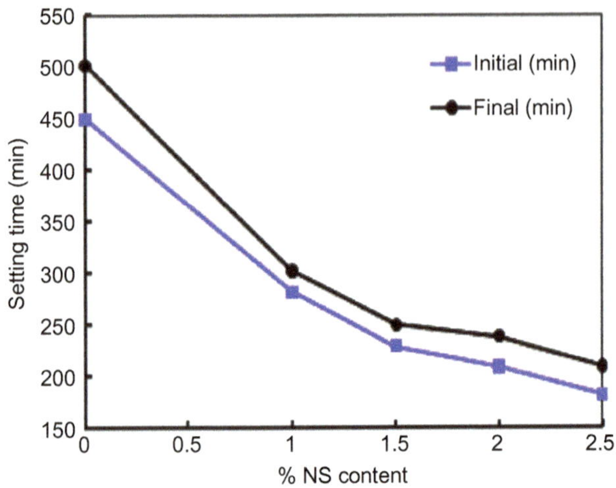

Fig. 4.5 Changes in initial and final setting times of mortar with nano-SiO_2 content [30].

workability tests, such as rheometric techniques, setting time, mini spread-flow, L-box (for self-consolidating concrete), and V funnel testing. It is reported that the addition of nano-SiO_2 can reduce the initial and final setting times of cement paste, mortar (Fig. 4.5), and concrete [28,30] because of the acceleration of hydration process. This applies to other types of 0-D nanoparticles. In addition, the initial and final setting times can decrease with the increase of nanoparticles. The magnitude of reduction depends on 0-D nanoparticle content, water-to-binder ratio (W/B), and mixture proportion. When 0-D nanoparticles are used in a small amount, they can fill voids among cement particles to replace some entrapped water. This can increase the fluidity of the mixture [31]. At a relatively high amount of 0-D nanoparticles, a large amount of free water and superplasticizer can be adsorbed onto the surface to increase the viscosity and eventually reduce the flowability [32]. Given the water-reducing admixture content, the loss of workability is proportional to the concentration of 0-D nanoparticles. This is because 0-D nanoparticles require more water to wet their surface due to high SSA, hence resulting in less workable and more viscous fresh mixture. Quercia et al. [33] found that fresh self-consolidating concrete with an addition of 3.8% nano-SiO_2, by mass of cement, showed similar flow properties and viscosity when compared to the reference mixture made without any nano-SiO_2. Increased torque, plastic viscosity, and yield stress are observed with the use of nano-SiO_2 and nano-TiO_2 in cement mortar [34]. It is important to mention that the increased viscosity can enhance the cohesiveness of nano-modified cement-based materials, which is beneficial in preventing segregation, especially for shotcrete applications [34].

Controversial findings regarding the effect of 0-D nanoparticles on the flowability of cement-based materials have been reported in the literature. This is because the workability is affected by W/B, mixture proportion, and main characteristics of 0-D nanoparticles, such as particle size distribution, shape, density, SSA, pore structure, and reactivity. The adverse effect on workability caused by 0-D nanoparticles can be alleviated by the use of SCMs, such as fly ash and slag, the increase in superplasticizer dosage, and the delayed addition of all the mixing water at a time [34,35].

4.5.2 Heat of hydration

Isothermal calorimetry can be used to evaluate the effect of 0-D nanomaterials on the early-age hydration of cement-based materials. Generally, the use of 0-D nanomaterials, such as nano-SiO_2, nano-TiO_2, nanoclay, and nano-$CaCO_3$, can accelerate the hydration process [9]. When 0-D nanoparticles are in contact with water, they can act as potential heterogeneous nucleation sites for the hydration products, and the grain boundary region can be densely populated with nuclei [36]. Thus, the intensity of the heat peak is increased, and the duration to reach the hydration peak of hydration and the dominant period are shortened, as shown in Fig. 4.6. The heat peak can be shifted from approximately 13 h to 8 h with the addition of nano-TiO_2. Besides, the cumulative heat released increased with the addition of 0-D nanoparticles. Worth mentioning is that nano-TiO_2 particles in the form of anatase show the potential to add new functionalities

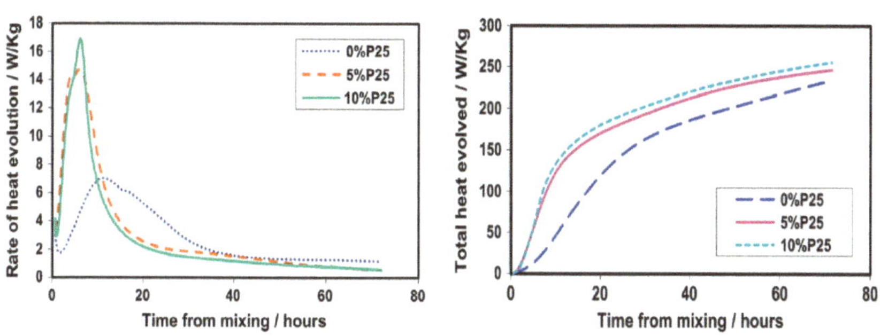

Fig. 4.6 Heat of hydration of paste made with different nano-TiO_2 contents. (A) Rate of heat evolution. (B) Total heat of hydration. (Note: P25 denotes mixture made with 75% anatase and 25% rutile) [37].

to infrastructures, such as self-cleaning capacity. However, the degree of hydration is slowed down after 7 days because most of the water available for hydration is absorbed by the 0-D nanoparticles [38].

4.5.3 Mechanical properties

The mechanical properties of concrete are the properties most valued by designers and quality control engineers. A lot of studies have been done on the effects of nano-SiO_2, nano-$CaCO_3$, and nano-iron on the mechanical properties of cement-based materials. Generally, with the increase of nanoparticles, the mechanical strength, including compressive, flexural, and tensile strengths, and modulus of elasticity are gradually increased. The pronounced effect of 0-D nanoparticles on strength is mainly focused at early ages before 7 d. Given the same diameter and SSA, it is reported that the inclusion of 0-D nano-SiO_2 increased the compressive strength of self-consolidating mortar more significantly at the early stages, when compared to those of nano-Al_2O_3 and nano-TiO_2 [39]. However, there is no significant difference among the mortars with various 0-D nanoparticles at the later ages of 28 and 90 d. Concrete made with nano-$CaCO_3$ is reported to have a greater effect on flexural strength when compared to compressive strength [24]. Meanwhile, the highest splitting tensile strength is observed on UHPC made with 3% nano-$CaCO_3$ compared to concrete made with the same amount of nano-SiO_2, nano-TiO_2, and nano-Al_2O_3 [40]. Some authors employed two or more types of 0-D nanoparticles in concrete and found that specimens containing combined nanoparticles can indeed show the highest improvement in mechanical properties [39].

A threshold limit of 0-D nanoparticles exists, beyond which the mechanical properties can be impaired [16]. This is because high 0-D nanoparticle content can easily result in agglomerate issues due to decreased workability, thus leading to more cracking and air voids near the embedded fibers and aggregate. Such cracks and air voids are triggers for the concrete failure, which can lead to reduced fiber-matrix bond properties and eventually decreased mechanical properties. Therefore, appropriate nanoparticles content should be used to secure dense microstructure and high strength and toughness. The optimal 0-D nanoparticles content varies in different ingredients, depending on various factors, such as composition, curing age, pozzolan type and content, chemical admixture, as well as nanoparticle size, shape, and type. Rong et al. [41] evaluated the changes in compressive and flexural strengths of UHPC made with nano-SiO_2 content, as illustrated

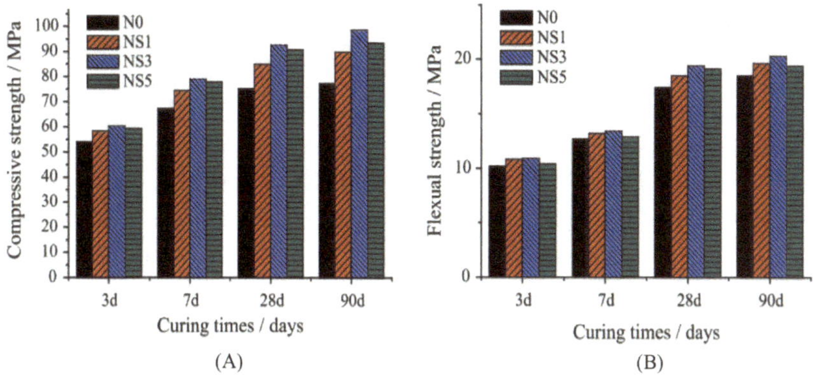

Fig. 4.7 Mechanical properties of UHPC made with different nano-SiO_2 contents. (A) Compressive strength. (B) Flexural strength. [41].

in Fig. 4.7. The highest compressive and flexural strengths are found at 3%, by mass of binder content. Wu et al. [16] found that the optimal nano-$CaCO_3$ and nano-SiO_2 contents for compressive and flexural strengths of UHPC matrix made with 20% silica fume at a 0.18 W/B are 1.6% to 4.8% and 0.5% to 1.5%, respectively. Ghafri et al. [42] investigated the effect of nano-SiO_2 addition (0, 1%, 2%, 3%, and 4%) on the mechanical properties of UHPC and found that the highest compressive strength is obtained at 3% nano-SiO_2, by mass of cement. Li et al. [43] reported threshold values of 1.0% and 3.0% for nano-SiO_2 and nano-$CaCO_3$, respectively, with respect to the mechanical strength of UHPC.

4.5.4 Shrinkage

Concrete shrinkage is due to the contraction of total mass upon loss of moisture and is expressed as a change in length per unit length. It is time dependent and can be categorized into plastic shrinkage, autogenous shrinkage, drying shrinkage, and carbonation shrinkage in terms of its cause. The primary concern concerning the shrinkage of concrete is the potential for cracking either in the plastic or hardened states [44]. Cracking can then serve as conduits for water and salt, hence, leading to reduced strength, durability, and service life of concrete [45].

It is reported that the drying shrinkage values increased with nano-SiO_{2}-content [46]. Replacing cement with nano-$CaCO_3$ increases the autogenous shrinkage of UHPC [43]. This is because the volume of hydration products formed is lower than that of the master materials. On the contrary, some researchers reported that the inclusion of 0-D nanoparticles, such as nano-SiO_2, nano-$CaCO_3$, nano-CaO, and nano-MgO, can decrease the drying or autogenous shrinkage of concrete [47,48]. A possible explanation is that the ultra-fine 0-D nanoparticles entrapped some water and then release it to help the hydration process. Such free water can increase the internal relative humidity, leading to lower capillary stresses according to Kelvin's law [49], and consequently reduce the autogenous shrinkage [48,50]. In addition, 0-D nanoparticles accelerate the hydration process, leading to faster development of internal microstructure, which acts as internal restraints against shrinkage. Furthermore, the addition of nanosize expansive agents, such as nano-CaO and nano-MgO, can form expansive products to compensate shrinkage of cement-based materials additionally. Zhang et al. [51] indicated that the addition of nano-TiO_2 can result in a comparable drying shrinkage to that of the control specimen at curing age of 1 month. This is because the decline of the contact angle and the refining of pore structure can slow down the water loss of hardened cement paste.

4.5.5 Durability

The diffusion of ions or gases into internal concrete is the main cause for concrete degradation, which includes carbonation, frost damage, leaching, sulfate attack, chloride intrusion, alkali-silica reaction, and steel corrosion. It is shown that concrete containing 0-D nanoparticles can exhibit lower water permeability and gas diffusivity compared to the control mixture without any 0-D nanoparticles [52,53]. The use of 0.3% nano-SiO_2 is reported to decrease the water permeability of concrete made with 0.48 W/B by 45% [54]. Najigivi et al. [55] evaluated two sizes of nano-SiO_2 (15 nm and 80 nm) on water absorption of concrete (W/B = 0.4) at the ages of 7, 28, and 90 d. Cement is partially replaced with nano-SiO_2 at levels of 0%, 1%, 1.5%, and 2%, by weight. The water absorption can decrease with the addition of 80 nm nano-SiO_2 at all ages. The addition of 1.5% led to the lowest percentage of water absorption at ages of 7 and 28 d, while the addition of 2% rendered the lowest percentage of water absorption at age of 90 d. The addition of nano-SiO_2 with 15 nm particle size increased the percentage of water absorption at age of 7 d, but decreased it at ages of 28 and 90 d.

The permeability to chloride migration can also decrease since the transport channels become partially blocked and disconnected [54]. However, the higher nano-SiO_2 content at 0.9% did not produce a proportionally higher resistance to chloride migration [54]. The nano-silica particle agglomerations are responsible for this compromised benefit on chloride resistance. Kong

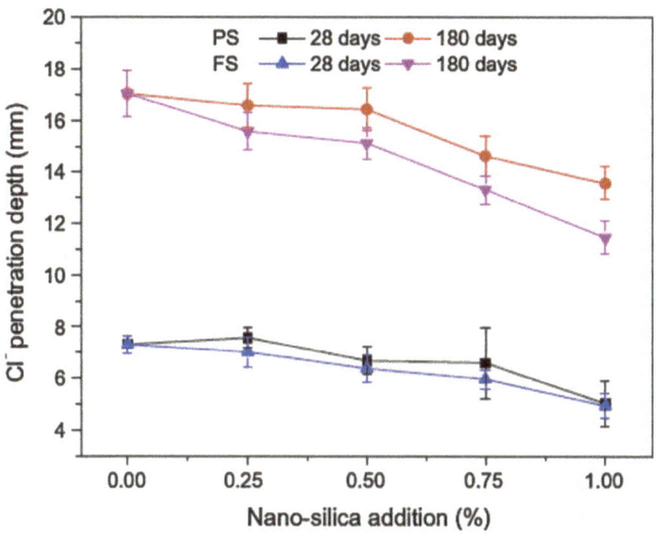

Fig. 4.8 Effect of nano-SiO_2 content on chloride penetration depth of mortar [56].

et al. [56] found that the chloride penetration depth of mortar can decrease gradually with nano-SiO_2 content increased from 0.25% to 1%, as illustrated in Fig. 4.8. Zhang and Li [57] reported that the resistance to chloride penetration of concretes with nano-TiO_2 can be greater than that with the same amount of nano-SiO_2. This is due to the smaller particle diameter and much greater SSA of nano-SiO_2 when compared to nano-TiO_2.

The addition of 0-D nanoparticles can increase the abrasion resistance and frost durability of concrete [33,58]. This is due to the reduction of the amount of $Ca(OH)_2$, especially at the weak ITZ. Some authors even found that 0-D nanoparticles are more favorable to the abrasion resistance of concrete than polypropylene fibers [59]. It is also found that the abrasion resistance of concrete containing nano-TiO_2 is better than that containing the same amount of nano-SiO_2 [59]. This is attributed to the finer particle diameter and greater SSA of nano-SiO_2. The addition of nano-SiO_2 into cement paste can significantly improve the thermal stability of the cementitious system. Compared to samples made with silica fume, nano-SiO_2-modified specimens showed less strength loss after exposure to an elevated temperature of 500°C [60].

4.5.6 Microstructure analysis

Microstructure analysis, such as porosity, micro and/nanomechanical properties, morphology of microstructure, of nano-reinforced cement-based materials is vital because of its close relationship to macro-performance, including mechanical properties, permeability, and visco-elastic properties. The microstructure of cement-based materials can be evaluated using SEM, energy dispersive X-ray analysis (EDX), XRD, mercury intrusion porosimetry (MIP), thermal-gravimetric (TG), microindention and nanoindention, and TEM techniques. Two or more techniques are often employed to reflect the microstructure of concrete. SEM can be utilized to reveal the morphology of hydration products, cement clinker, and pores. EDX is usually used with SEM together. MIP is an efficient technique for reflecting the porosity and pore size distribution of cement-based materials. Lower porosity means higher strength and durability. XRD can provide qualitative and quantitative information on crystal substances presented in cement-based materials, which include SiO_2, portlandite, C_3S, ettringite, calcite, and even some C—S—H, through identifying corresponding intensity peaks. TEM can be employed to spatially determine the chemical compositions of the main hydration products of C—S—H in cement paste. Microindention and nanoindention can be used to determine local mechanical properties, such as elastic modulus and hardness. The short-term creep behaviors of C—S—H phases can be detected using the nanoindentation test.

SEM images showed that the 0-D nanoparticles are acting as fillers and serving as nucleation seeds to promote the hydration process and improve the microstructure of cement paste if they are uniformly dispersed [61]. This can render a more uniform and denser microstructure of nano-modified cement-based materials than that of the control mixture. The nano-SiO_2 is also found to reduce the quantity and size of calcium hydroxide crystals and form secondary C—S—H at the interface due to the pozzolanic effect, based on TG and XRD analyses [41]. The reduction of mass loss at temperature range of 400°C to 500°C implies the consumption of calcium hydroxide. This can produce a more homogeneous microstructure [54,62], as shown in Fig. 4.9. Calcium hydroxide crystals can have a detrimental effect on the performance of concrete due to leaching, carbonation, and introduction of air voids and porous interface because of their preferential orientation to the surface of the fiber and/or aggregate [63].

Fig. 4.10 illustrates MIP results of the effect of nano-SiO_2 content (0%, 0.5%, 1.5%, and 2.5%) on porosity and pore size distribution of mortar. With the increase of nano-SiO_2 content, the porosity decreased with nano-SiO_2 content increased from 0% to 1.25% and then decreased with a further addition to 2.5%. The MIP results agree well with the mechanical properties. The formation of C—S—H and other hydrates related to the suitable

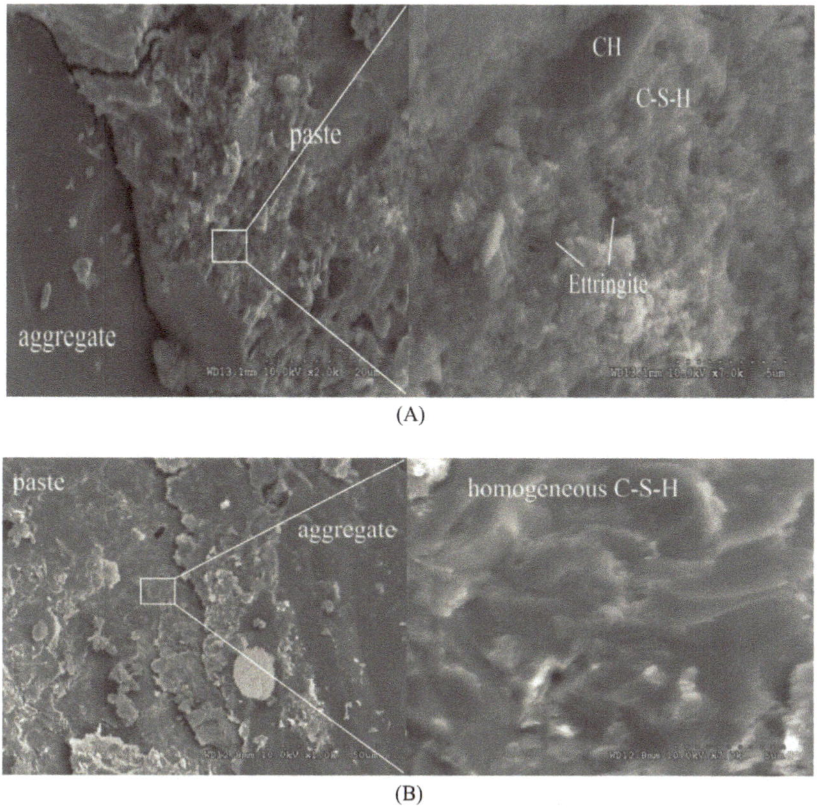

Fig. 4.9 Microstructure morphologies at the ITZ. (A) Ordinary portland concrete. (B) Concrete with 0.9% nano-SiO_2 [54].

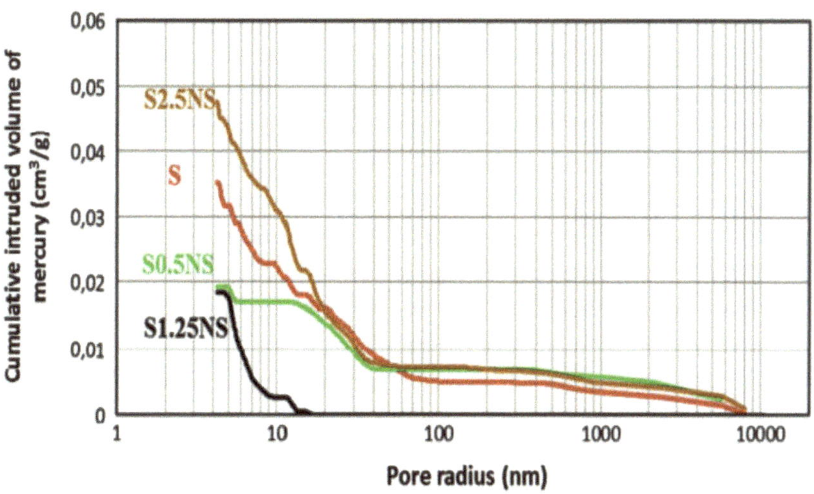

Fig. 4.10 MIP results of mortar with different contents of nano-SiO_2 [64].

use of 0-D nanoparticles content may decrease the total porosity, refine the size distribution of pores, and reduce the pore connectivity, as reflected from MIP analysis. The reduction of porosity mainly occurred within the capillary pore range. As permeability and penetration of harmful substances into concrete are affected mainly by the large and medium capillary pores, the decrease in those pore ranges is a good indication of enhanced durability [65]. Meanwhile, the critical pore size of concrete with the addition of nano-SiO_2 is reduced. Critical pore diameter presents a mean size of pore entryways that allows maximum percolation throughout the pore system. The increased porosity associated with the high content of 0-D nanoparticles is attributed to entrapped air voids associated with agglomeration issues. Microhardness tests indicated that the microhardness of cement-based materials made with appropriate 0-D nanoparticles can be increased. From the nanoindentation studies, it is observed that the nano-SiO_2 can significantly increase the high stiffness of C—S—H which is characterized by a lower Ca/Si ratio [35].

4.6 Conclusions and outlook

The use of 0-D nanomaterials in cementitious materials has been extensively studied in the laboratory in terms of their effect on fresh and hardened properties. The addition of 0-D nanoparticles in paste, mortar, and concrete is a good way to enhance their properties and develop high-performance concrete and UHPC. Based on the contribution of nanoparticles in cement-based materials, the following conclusions can be drawn:

(1) The addition of ultra-fine and high-specific surface 0-D nanoparticles into cement-based materials can increase the water and water-reducing agent amount and shorter the setting time. Thus, the workability of the cement-based materials can reduce with the increase of 0-D nanoparticles content. Meanwhile, 0-D nanoparticles can act as potential nucleation sites for hydration products to accelerate the hydration process.

(2) The use of suitable content of 0-D nanoparticles can contribute to a dramatic increase in the mechanical strength of cementitious materials, thus helping the production of HPC and UHPC. The mechanical strength can increase with the addition of 0-D nanoparticles up to a specified limit and then decrease attributed to agglomerations issues. The optimum 0-D nanoparticle content is still controversial and depends on many factors, such as W/B, mixture proportion, size, and type of 0-D nanoparticles. The related mechanisms to enhance mechanical properties are as follows:

- Filling the spaces and pores between the coarse and fine particles to render close packing of the solid system.
- Acting as nucleation sites for C—S—H precipitation, thus contributing to the development of hydration in Portland cement and forming more hydration products.
- Chemical reaction with $Ca(OH)_2$ crystals and/or C_3A to produce C—S—H gel and reduce the quantity and size of $Ca(OH)_2$, especially for nano-SiO_2 and nano-$CaCO_3$.

(3) 0-D nanoparticles exert a great influence on enhancing the durability of cement-based materials due to their chemical and physical properties.

(4) Appropriate use of 0-D nanoparticles in cement-based materials can greatly densify and homogenize the microstructure and probably induce the small-sized C—S—H gel precipitation with higher stiffness and lower Ca/Si ratio. The quality of the interface between bulk cementations matrix and aggregate/fiber can be improved to enhance mechanical properties and durability. However, weak zones around aggregate or fibers can be introduced to increase porosity due to agglomeration issues, especially at relatively high content of the nanoparticles.

Current challenges need to be solved before realizing the full potential of nanotechnology in concrete applications, including proper dispersion, compatibility of the 0-D nanomaterials with other components, processing and manufacturing, safety, handling issues, and material cost.

References

[1] L. Sorelli, G. Constantinides, F.J. Ulm, F. Toutlemonde, The nano-mechanical signature of ultra high performance concrete by statistical nanoindentation techniques, Cem. Concr. Res. 38 (12) (2008) 1447–1456.

[2] H.M. Jennings, J.W. Bullard, J.J. Thomas, J.E. Andrade, J.J. Chen, G.W. Scherer, Characterization and modeling of ores and surfaces in cement paste: correlations to processing and properties, J. Adv. Concrete Technol. 6 (2008) 5–29.

[3] H. Bahadori, P. Hosseini, Reduction of cement consumption by the aid of silica nanoparticles (investigation on concrete properties), J. Civ. Eng. Manag. 18 (3) (2012) 416.

[4] F. Ulm, M. Vandamme, Probing nano-structure of C–S–H by micro-mechanics based indentation techniques, in: Z. Bittnar, P.J.M. Bartos, J. Nemecek, V. Smilauer, J. Zeman (Eds.), Nanotechnology in Construction: Proceedings of the NICOM3 (3rd International Symposium on Nanotechnology in Construction). Prague, Czech Republic, 2009, pp. 43–53.

[5] M.J. DeJong, F.J. Ulm, The nanogranular behavior of C–S–H at elevated temperatures (up to 700°C), Cem. Concr. Res. 37 (1) (2007) 1–12.

[6] P. Mondal, S.P. Shah, L. Marks, Nanoscale characterization of cementitious materials, ACI Mater. J. 105 (2008) 174–179.
[7] M. Wilson, K.K.G. Smith, M. Simmons, B. Raguse, Nanotechnology—Basic Science and Emerging Technologies, Chapman & Hall/CRC, 2000.
[8] K. Sobolev, M. Ferrada-Gutiérrez, How nanotechnology can change the concrete world: part 2, Am. Ceram. Soc. Bull. 11 (2005) 16–19.
[9] B.W. Jo, C.H. Kim, G.H. Tae, J.B. Park, Characteristics of cement mortar with nano-SiO_2 particles, Construct. Build Mater. 21 (6) (2007) 1351–1355.
[10] T. Phoo-ngernkham, P. Chindaprasirt, V. Sata, S. Hanjitsuwan, S. Hatanaka, The effect of adding nano-SiO_2 and nano-Al_2O_3 on properties of high calcium fly ash geopolymer cured at ambient temperature, Mater. Des. 55 (2014) 58–65.
[11] T. Meng, Y. Yu, Z. Wang, Effect of nano-$CaCO_3$ slurry on the mechanical properties and micro-structure of concrete with and without fly ash, Compos. Part B Eng. 117 (2017) 124–129.
[12] A. Khoshakhlagh, A. Nazari, G. Khalaj, Effects of Fe_2O_3 nanoparticles on water permeability and strength assessments of high strength self-compacting concrete, J. Mater. Sci. Technol. 28 (1) (2012) 73–82.
[13] P. Sikora, E. Horszczaruk, K. Cendrowski, E. Mijowska, The influence of nano-Fe_3O_4 on the microstructure and mechanical properties of cementitious composites, Nanoscale Res. Lett. 11 (1) (2016) 182.
[14] T. Meng, Y. Yu, X. Qian, S. Zhan, K. Qian, Effect of nano-TiO_2 on the mechanical properties of cement mortar, Construct. Build Mater. 29 (2012) 241–245.
[15] L.L. Zhai, G.P. Ling, Y.W. Wang, Effect of nano-Al_2O_3 on adhesion strength of epoxy adhesive and steel, Int. J. Adhes. Adhes. 28 (1–2) (2008) 23–28.
[16] Z. Wu, C. Shi, K.H. Khayat, S. Wan, Effects of different nanomaterials on hardening and performance of ultra-high strength concrete (UHSC), Cem. Concr. Compos. 70 (2016) 24–34.
[17] S. Kawashima, P. Hou, D.J. Corr, S.P. Shah, Modification of cement-based materials with nanoparticles, Cem. Concr. Compos. 36 (2013) 8–15.
[18] J. Gaitero, I. Campillo, A. Guerrero, Reduction of the calcium leaching rate of cement paste by addition of silica nanoparticles, Cem. Concr. Compos. 38 (8) (2008) 1112–1118.
[19] A. Najigivi, A. Khaloo, S.A. Rashid, Investigating the effects of using different types of SiO_2 nanoparticles on the mechanical properties of binary blended concrete, Compos. Part B Eng. 54 (2013) 52–58.
[20] L.Y. Yang, Z.J. Jia, Y.M. Zhang, J.G. Dai, Effects of nano-TiO_2 on strength, shrinkage and microstructure of alkali activated slag pastes, Cem. Concr. Compos. 57 (2015) 1–7.
[21] M.H. Zhang, H. Li, Pore structure and chloride permeability of concrete containing nanoparticles for pavement, Construct. Build Mater. 25 (2011) 608–616.
[22] N. Mandzy, E. Grulke, T. Druffel, Breakage of TiO_2 agglomerates in electrostatically stabilized aqueous dispersions, Powder Technol. 160 (2) (2005) 121–126.
[23] K. Sobolev, I. Flores, L.M. Torres-Martinez, P.L. Valdez, E. Zarazua, E.L. Cuellar, Engineering of SiO_2 Nanoparticles for Optimal Performance in Nano Cement Based Materials, Springer, Berlin Heidelberg, 2009, pp. 139–148.
[24] Z. Wu, C. Shi, K.H. Khayat, Multi-scale investigation of microstructure, fiber pullout behavior, and mechanical properties of ultra-high performance concrete with nano-$CaCO_3$ particles, Cem. Concr. Compos. 86 (2018) 255–265.
[25] N.T. Thanh, N. Maclean, S. Mahiddine, Mechanisms of nucleation and growth of nanoparticles in solution, Chem. Rev. 114 (15) (2014) 7610–7630.
[26] F.J. Tang, E.M. Gartner, Influence of sulfate source on Portland cement hydration, Adv. Cem. Res. 1 (2) (1988) 67–74.

[27] G. Land, D. Stephan, Controlling cement hydration with nanoparticles, Cem. Concr. Compos. 57 (2015) 64–67.
[28] M.H. Zhang, J. Islam, S. Peethamparan, Use of nano-silica to increase early strength and reduce setting time of concretes with high volumes of slag, Cem. Concr. Compos. 34 (5) (2012) 650–662.
[29] K. De Weerdt, M.B. Haha, G.L. Saout, H. Justnesa, B. Lothenbach, Hydration mechanisms of ternary Portland cements containing limestone powder and fly ash, Cem. Concr. Res. 41 (3) (2011) 279–291.
[30] L. Senff, J.A. Labrincha, V.M. Ferreira, D. Hotza, W.L. Repette, Effect of nano-silica on rheology and fresh properties of cement pastes and mortars, Construct. Build Mater. 23 (2009) 2487–2491.
[31] J. Liu, Q. Li, S. Xu, Influence of nanoparticles on fluidity and mechanical properties of cement mortar, Construct. Build Mater. 101 (2015) 892–901.
[32] P. Hou, S. Kawashima, K. Wang, D.J. Corr, J. Qian, S.P. Shah, Effects of colloidal nanosilica on rheological and mechanical properties of fly ash-cement mortar, Cem. Concr. Compos. 35 (2013) 12–22.
[33] G. Quercia, P. Spiesz, H.J.H. Brouwers, Effects of nano-silica (NS) additions on durability of SCC mixtures, in: Durability of Reinforced Concrete from Composition to Protection, Springer International Publishing, 2015, pp. 125–143.
[34] M. Berra, F. Carassiti, T. Mangialardi, A.E. Paolini, M. Sebastiani, Effects of nanosilica addition on workability and compressive strength of Portland cement pastes, Construct. Build Mater. 35 (2012) 666–675.
[35] P. Hou, S. Kawashima, D. Kong, D.J. Corr, J. Qian, S.P. Shah, Modification effects of colloidal nanoSiO_2 on cement hydration and its gel property, Compos. Part B Eng. 45 (2013) 440–448.
[36] J.J. Thomas, A new approach to modeling the nucleation and growth kinetics of tricalcium silicate hydration, J. Am. Ceram. Soc. 90 (10) (2007) 3282–3288.
[37] J. Chen, S. Kou, C.S. Poon, Hydration and properties of nano-TiO_2 blended cement composites, Cem. Concr. Compos. 34 (2012) 642–649.
[38] H. Madani, A. Bagheri, T. Parhizkar, The pozzolanic reactivity of monodispersed nanosilica hydrosols and their influence on the hydration characteristics of Portland cement, Cem. Concr. Res. 42 (12) (2012) 1563–1570.
[39] E. Mohseni, B.M. Miyandehi, J. Yang, M.A. Yazdi, Single and combined effects of nano-SiO_2, nano-Al_2O_3 and nano-TiO_2 on the mechanical, rheological and durability properties of self-compacting mortar containing fly ash, Construct. Build Mater. 84 (2015) 331–340.
[40] S. Yu, J. Li, C. Wu, P. Wu, Z.X. Li, Influences of nano-particles on dynamic strength of ultra-high performance concrete, Compos. Part B Eng. 91 (2016) 595–609.
[41] Z.D. Rong, W. Sun, H.J. Xiao, G. Jiang, Effects of nano-SiO_2 particles on the mechanical and microstructural properties of ultra-high performance cementitious composites, Cem. Concr. Compos. 56 (2015) 25–31.
[42] E. Ghafari, H. Costa, E. Júlio, A. Portugal, L. Durães, The effect of nanosilica addition on flowability, strength and transport properties of ultra high performance concrete, Mater. Des. 59 (2014) 1–9.
[43] W. Li, Z. Huang, T. Zu, C. Shi, W.H. Duan, S.P. Shah, Influence of nanolimestone on the hydration, mechanical strength, and autogenous shrinkage of ultrahigh-performance concrete, J. Mater. Civ. Eng. 28 (1) (2015) 04015068.
[44] G.A. Rao, Long-term drying shrinkage of mortar-influence of silica fume and size of fine aggregate, Cem. Concr. Res. 31 (2001) 171–175.
[45] H.A. Toutanji, Properties of polypropylene fiber reinforced silica fume expansive-cement concrete, Construct. Build Mater. 13 (1999) 171–177.

[46] A. Sadrmomtazi, A. Fasihi, The role of Nano-SiO_2 in mechanical properties of rice husk ash composite cement mortars, Nanotechnol. Nanosci. 2 (1) (2011) 42–45.
[47] J. Camiletti, A.M. Soliman, M.L. Nehdi, Effects of nano-and micro-limestone addition on early –age properties of ultra-high-performance concrete, Mater. Struct. 46 (2013) 881–898.
[48] R. Polat, R. Demirboğa, W.H. Khushefati, Effects of nano and micro size of CaO and MgO, nano-clay and expanded perlite aggregate on the autogenous shrinkage of mortar, Construct. Build Mater. 81 (2015) 268–275.
[49] O. Coussy, P. Dangla, T. Lassabate're, V. Baroghel-Bouny, The equivalent pore pressure and the swelling and shrinkage of cement-based materials, Mater. Struct. 37 (1) (2004) 15–20.
[50] E. Holt, Contribution of mixture design to chemical and autogenous shrinkage of concrete at early ages, Cem. Concr. Res. 35 (3) (2005) 464–472.
[51] R. Zhang, X. Cheng, P. Hou, Z. Ye, Influences of nano-TiO_2 on the properties of cement-based materials: hydration and drying shrinkage, Construct. Build Mater. 81 (2015) 35–41.
[52] T. Ji, Preliminary study on the water permeability and microstructure of concrete incorporating nano-SiO_2, Cem. Concr. Res. 35 (2005) 1943–1947.
[53] T.P. Chang, J.Y. Shih, K.M. Yang, T.C. Hsiao, Material properties of Portland cement paste with nano-montmorillonite, J. Mater. Sci. 42 (17) (2007) 7478–7487.
[54] H. Du, S. Du, X. Liu, Durability performances of concrete with nano-silica, Construct. Build Mater. 73 (2014) 705–712.
[55] A. Najigivi, S.A. Rashid, F.N.A. Aziz, M.A.M. Sakkeh, Water absorption control of ternary blended concrete with nano-SiO_2 presence of rice husk ash, Mater. Struct. 45 (2012) 1007–1017.
[56] D. Kong, X. Du, W. Su, H. Zhang, Y. Yang, S.P. Shah, Influence of nano-silica agglomeration on microstructure and properties of the hardened cement-based materials, Construct. Build Mater. 37 (2012) 707–715.
[57] M.H. Zhang, H. Li, Pore structure and chloride permeability of concrete containing nano-particles for pavement, Construct. Build Mater. 25 (2011) 608–616.
[58] K. Ibrahim Rahel, R. Hamid, M.R. Taha, Fire resistance of high-volume fly ash mortars with nanosilica addition, Construct. Build Mater. 36 (2012) 779–786.
[59] H. Li, M.H. Zhang, J.P. Ou, Abrasion resistance of concrete containing nanoparticles for pavement, Wear 260 (2006) 1262–1266.
[60] S. Lim, P. Mondal, I. Cohn, Effects of nanosilica on thermal degradation of cement paste, in: NICOM 4—4th International Symposium on Nanotechnology in Construction, Greece, 2012.
[61] H. Li, H.G. Xiao, J. Yuan, J. Ou, Microstructure of cement mortar with nano particles, Compos. Part B Eng. 35 (2004) 185–189.
[62] Q. Ye, Z. Zhang, D. Kong, R. Chen, Influence of nano-SiO_2 addition on properties of hardened cement paste as compared with silica fume, Construct. Build Mater. 21 (2007) 539–545.
[63] T. Akçaoğlu, M. Tokyay, T. Çelik, Assessing the ITZ microcracking via scanning electron microscope and its effect on the failure behavior of concrete, Cem. Concr. Res. 35 (2) (2005) 358–363.
[64] M. Oltulu, R. Sahin, Pore structure analysis of hardened cement mortars containing silica fume and different nano-powders, Construct. Build Mater. 53 (2014) 658–664.
[65] A.M. Neville, Properties of Concrete, fourth ed., Prentice Hall/Pearson, Harlow, 2002, pp. 537–576.

CHAPTER 5

Design, mechanism, and performance of cement-based materials with 1D nanomaterials

Weina Meng[a] and Kamal H. Khayat[b]
[a]Department of Civil, Environmental and Ocean Engineering, Stevens Institute of Technology, Hoboken, NJ, United States
[b]Center for Infrastructure Engineering Studies, Department of Civil, Architectural and Environmental Engineering, Missouri University of Science and Technology, Rolla, MO, United States

5.1 Introduction

The major disadvantage of cement-based materials is their brittle nature which is attributed to their poor resistance to crack formation, the low tensile strength, and low tensile strain capacities. Efforts have been conducted to enhance the tensile properties of cement-based materials by adding micro or/and macro-fibers. Fibers change large wide cracks to dense microcracks. However, the fibers fail to stop cracking initiation at nanoscale and propagation at microscale. Extensive research endeavors over the last few years demonstrated the application potential of various 1D nanomaterials, mainly carbon nanofiber (CNF) and carbon nanotube (CNT), in cementitious composites. Compared with 0D nanomaterials, 1D nanomaterials can effectively restrict the formation and growth of nanocracks in cementitious composites, which benefits the development of high crack-resistant cement-based materials with superior mechanical properties and durability. This chapter reviews recent research on design, performance, and mechanism of cement-based materials with 1D nanomaterials.

5.2 Types and characteristics of 1D nanomaterials

5.2.1 Carbon nanotubes

The CNTs are a class of 1D nanomaterials with a cylindrical nanostructure which consist of curved graphene layer made of carbon atoms in a honeycomb structure. The CNTs are divided into single-walled CNTs and multi-walled CNTs (MWCNTs) based on the number of graphene layers used in their structure [1] (Fig. 5.1). The CNTs have excellent mechanical, thermal,

Nanotechnology for Civil Infrastructure
https://doi.org/10.1016/B978-0-12-817832-4.00004-3

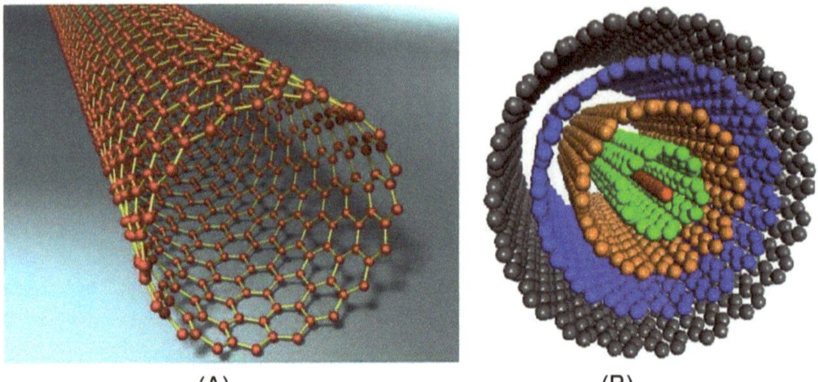

Fig. 5.1 Nanostructure of (A) single-walled CNTs and (B) multiwalled CNTs [2].

electrical, and chemical properties [3–8]. In addition, the unique structure of CNTs provides them with a variety of applications [9–12]. Besides the chemical composition, the geometry of CNTs has also shown an influencing factor on the performance of CNTs' reinforced cement-based materials [2,13]. The CNTs offer substantial aspect ratios for improving the mechanical properties of cement-based materials by promoting the hydration reactivity.

5.2.1.1 Properties of carbon nanotube

The CNTs have distinct tensile properties. The CNTs, as the stiffest fibers, Young's Modulus is reported in the range of 1–1.4 TPa [14,15], the elongation to failure is 20%–30%, and the tensile strength ranging from 11 to 63 GPa [16]. The superior mechanical properties of the CNTs are due to the strong bonding between carbon atoms [16]. However, weak shear interaction between shells and tubes limits the applicability of CNTs [17]. Nevertheless, this problem can be solved by using high-energy electron irradiation, which can significantly strengthen the shear by creating strong crosslinks between shells and tubes [18].

As a functional material, the CNTs also exhibit extraordinary thermal conductivity. For nonmetallic materials, heat is mainly carried by phonons that are defined as the quanta of crystal lattice vibration [19]. Phonon conduction mechanism is dominated by phonon active modes, the boundary surface scattering, and the length of the free path for the phonons [20]. Therefore, quantitative measurement of thermal conductivity is challenging. Researchers have alternatively employed theoretical simulations and in-direct experimental tools to study the thermal performance of CNTs. Obtained results indicated that the thermal conductivity of CNTs varies from 2000 to 6000 W/mK [21–23].

In addition, the CNTs have unique electrical properties which mainly stem from their 1D geometry and the electronic structure of graphite [24]. Small diameter and high length-to-diameter ratio of CNTs prevent the deflection of an electron from its path and thereby result in low electrical resistance of CNTs [16,25]. Further, the high current density facilitates electron transition within CNTs, which in turn reduces the electrical resistivity [26,27]. It is reported that the electrical resistance of CNTs is in the range of 5.1×10^{-6} to 1.2×10^{-4} Ωcm [28,29].

5.2.2 Carbon nanofibers

More straightforward synthesizing processes along with lower prices are the most important advantages of CNFs over the CNTs. Therefore, the application of CNFs in cement composites attracts more and more interests. The CNFs have a diameter in the range of 50–500 nm and a high aspect ratio (typically more than 50 m^2/g) [30–32]. CNFs are typically made of a single or double layer of graphite planes that can be stacked parallel or at a certain angle from the fiber axis [33,34] (Fig. 5.2). It has been shown that the manufacturing process and the posttreatment methods could affect the properties of CNFs [35,36]. The CNFs have a Young's Modulus as high as 400 GPa, with a tensile strength of 7–30 GPa [37]. The CNFs have shown

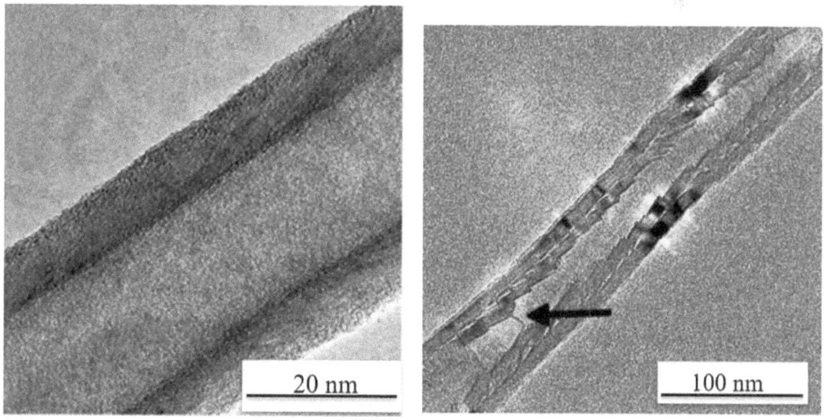

(A) (B)

Fig. 5.2 TEM micrograph of CNFs showing (A) a single layer and (B) double layers of graphite planes [35,36].

to have lower thermal conductivity than CNTs, which has reported to be 1900 W/mK [38]. The electrical resistance for CNFs has reported to be 1000 Ω cm [38].

5.3 Dispersing of 1D nanomaterials in cementitious matrices

When adding nanomaterials into the cement matrix and solvent, the distribution of nanomaterials has a significant influence on the properties of nanocomposites [39]. The CNTs and CNFs, as the most commonly used 1D nanomaterials in cementitious composites, suffer from poor dispersion [38]. Strong van der Waals force in CNTs and CNFs is the primary factor causing agglomeration of CNFs and CNTs in cementitious composites [40]. Large specific surface area, large aspect ratio and volume fraction, and long length (i.e., causes interlocking) of the 1D nanomaterials, as well as low viscosity of the cementitious matrix suspension are other important factors that influence the tendency of these nanomaterials to agglomerate, which in turn affects the uniform dispersion of CNFs and CNTs [38–40].

Besides undesirable effect on the performance of cementitious composite, poor dispersion of nanomaterials can adversely affect the hydration process of cement and strength development. Stress concentration in vicinity of the agglomerated area can generate local cracks which negatively affect the performance of cementitious composite [38,41]. In this regard, uniformity of CNTs and CNFs dispersed in the cementitious matrix is a critical factor for the application of these nanomaterials in cementitious composite. Consequently, researchers have put considerable effort into finding methods to prevent agglomeration or to split the formed agglomerates [39–41].

5.3.1 Physical dispersion techniques

5.3.1.1 Ultrasonication

Ultrasonic method, [42–55], ball milling, and mechanical stirring methods [56,57] are methods used to disperse nanomaterials within a cementitious environment. Compared to the ultrasonic method, mechanical stirring and ball milling methods have shown to be less effective. That is why these methods are usually used as preliminary treatment for the suspension of CNTs [58,59].

Ultrasonication is a method in which a voltage applied to a probe generates mechanical vibrations. As a result of these vibrations, pressure waves are generated in the fluid, and bubbles are created on a microscale in the

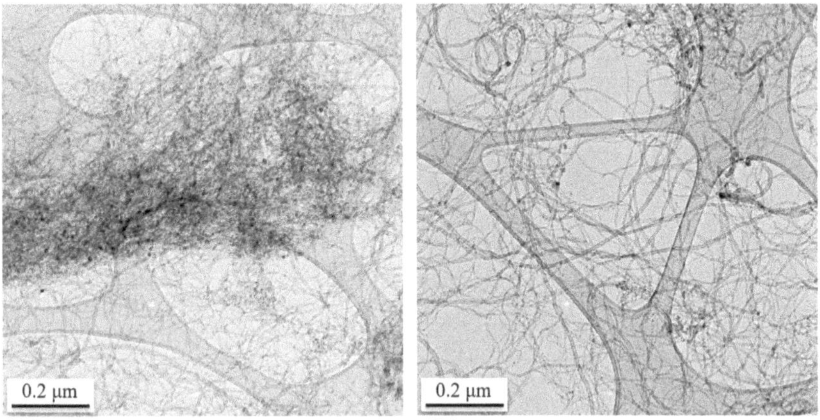

(A) (B)

Fig. 5.3 SEM images of CNT: (A) before sonication and (B) after sonication [38].

solution. Through a process known as cavitation, bubbles collapse and release a high level of energy. The released energy leads to the dispersion of nanoparticles [41,51,60,61]. The ultrasonic equipment utilizes two methods for CNTs and CNFs dispersion: ultrasonic bath and ultrasonic tip [49,62,63]. The ultrasonic tip has shown to be more effective in the dispersion of CNTs than the ultrasonic bath as it generates denser energy with lower frequency [64]. Fig. 5.3 shows the effectiveness of the ultrasonic method in the dispersing of CNTs [38].

Although the ultrasonic method has been shown to be highly efficient in water and other liquid with low viscosity, this method cannot guarantee the uniform dispersion of CNTs and CNFs in the cement environment [39]. For this reason, the ultrasonic method is mostly used with other dispersion methods [47,65–67]. For instance, Konsta-Gdoutos et al. [49] used the combination of ultrasonication and surfactant methods for the dispersion of CNTs in cementitious composites. They disclosed that for proper dispersion of CNTs, at least 70 Pa of ultrasonic energy was needed. Some researchers [68,69] reported that ultrasonication energy and time had a great influence on the CNT dispersion. In this regard, long duration with high energy ultrasonication was reported to damage CNTs [68].

5.3.2 Chemical methods

5.3.2.1 Use of surfactant

Surfactants are a group of the compound known for their amphiphilic structure which means that their molecules have a functional group called the

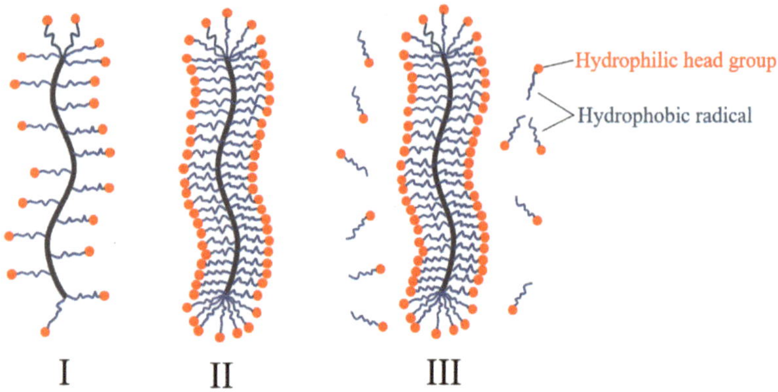

Fig. 5.4 Schematic view of the adsorbed surfactant molecules on the CNT surface: I—low surfactant concentration; II—the condition of equilibrium, that is, the formation of rod-like micelles; III—oversaturated surfactant concentration [70].

hydrophilic group that has an affinity to water, and at the same time, they have a functional group called a hydrophobic group, which does not show affinity to water [41].

The effect of surfactant concentration and the mechanism through which a surfactant improves the dispersion of nanomaterials was investigated by Sobolkina et al. [70]. As shown in Fig. 5.4, the hydrophobic groups of the surfactant adsorbed on the surface of CNTs and formed a micelle in which the hydrophobic group faces inward, while the hydrophilic group faces outward. The surfactant improves the dispersion of CNTs by means of electrostatic and/or steric repulsions between the surfactant molecules. As can be seen from Fig. 5.4, in low concentration, the surfactant was not able to adequately coat the surface of CNTs. In this case, there are still strong surface effects between the CNTs. At an optimum concentration, the surface of the CNT was entirely covered with surfactant molecules which hinder the agglomeration of CNTs. It was also reported that too much surfactants could not improve the CNTs' dispersion but may cause agglomeration [71]. Finding the optimum concentration is key for the uniform dispersion.

However, some research revealed that some types of surfactants might not be compatible with the cement matrix [72,73]. In this regard, selecting the appropriate surfactant is also a critical factor that significantly affects the efficiency of CNT's and CNF's dispersion in the cementitious systems. For instance, Ref. [72] disclosed that some surfactants such as Gum Arabic could have an adverse influence on the dispersion of nanomaterials by reacting with some superplasticizer. It has also been shown that surfactants that are incompatible with the cement matrix could have improper effects such as preventing hydration and entrap substantial air in the cementitious materials [72]. On the other hand, the surfactants such as nonionic polyoxythylene laurylether [70], methylcellulose [74], and pluronic F-127 [75] are reported as proper surfactant that are able to effectively disperse CNTs' particles in cement matrix without affecting the hydration process.

5.3.2.2 Use of typical cement admixtures

Recently, the influence of admixtures typically used in cement on the dispersion of CNTs and CNFs in cementitious systems has been intensively investigated. Collins et al. [73] compared the capability of air entrainer, lignosulfonate, and polycarboxylate-type superplasticizer, as cement admixtures, in dispersing CNTs in Portland cement mixtures. Among the selected admixtures, polycarboxylates were found to have better performance. They noted that the superiority of polycarboxylates was because of the electrostatic repulsion caused by the carboxylate groups and the steric repulsion due to the long lateral ether chains. Fig. 5.5 shows the SEM picture of 28 days CNT-OPC hardened paste samples. As is clear, adding

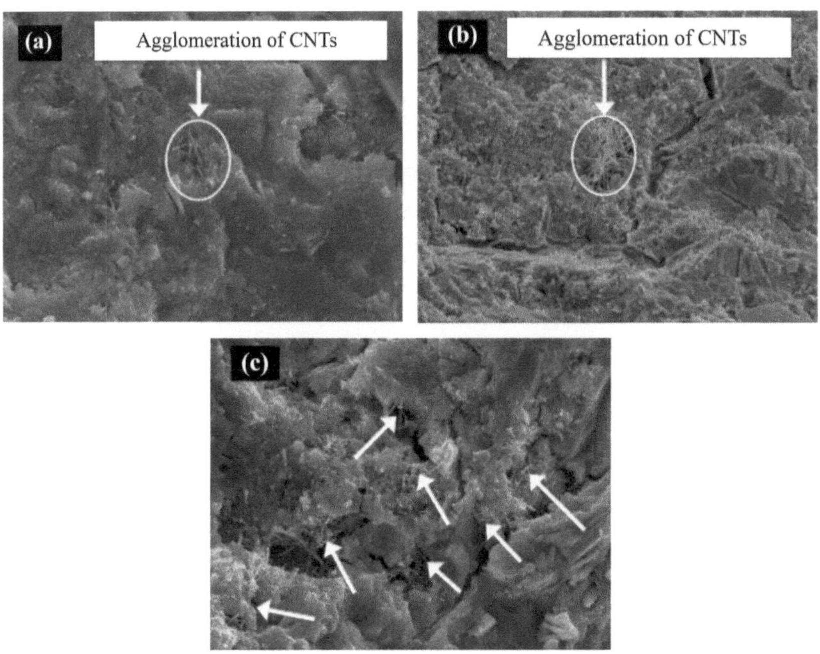

Fig. 5.5 Effect of admixtures on dispersion of CNTs: (A) no admixture, (B) air entrainer, and (C) polycarboxylate [73].

polycarboxylates significantly improved the dispersion of CNTs in the cement matrix. Kim et al. [76] investigated the combined effect of pozzolanic materials and superplasticizers on the dispersion of CNTs in the cement matrix. Specifically, they used silica fume and nanosilica as a pozzolanic material, and lignosulfonate, naphthalene, and polycarboxylate as superplasticizer. They revealed that adding silica fume together with superplasticizer could further enhance the dispersion of CNTs in cementitious composites. However, it was shown that adding nano-silica was not effective in the dispersion of CNTs.

5.3.2.3 Covalent functionalization

The covalent functionalization method is to synthesize a covalent linkage between the functional group and the carbon skeleton of CNTs or CNFs. The covalent linkage allows the hydrophilic functional groups attached to the surface of CNTs and CNFs and eventually improves the dispersion of CNTs and CNFs by reducing the contact angle and increasing the wettability of CNTs and CNFs when contact with water [42]. In addition, it has been shown that the synthesized functional groups on the surface of CNTs and CNFs can react with calcium silicate hydrate (C—S—H) and calcium hydroxide (C—H) gel and form a strong covalent linkage within the cement matrix [74].

Recent researches reveal that the acid treatment is an efficient method to create functional groups on the surface of CNTs and CNFs in the cement matrix [42,74,77]. Li et al. [74] showed that utilizing strong acids such as H_2SO_4 and HNO_3 formed carboxylic acid groups on the surface of CNT, which significantly improved the dispersion of CNTs in cementitious composites.

Covalent synthesizing of pozzolanic materials in order to improve the dispersion of CNTs and CNFs has also been studied by researchers [78,79]. For instance, Kim et al. [79] used silica fume as a pozzolanic material and functionalized the surface of silica fume with a primary amine group and the surface of CNTs with carbonyl chloride group in order to conjugate the CNTs and silica fume. They reported that the amine group and carbonyl chloride group caused silica fume and CNTs to form a covalent bond (Fig. 5.6), which improved the dispersion of CNTs in the cement matrix.

5.3.2.4 Combination of various methods

As stated earlier, each of the above-mentioned methods has its own merits and demerits. Therefore, it is hardly possible to achieve the optimal dispersing of CNTs and CNFs, using only one method. In this regard, researchers

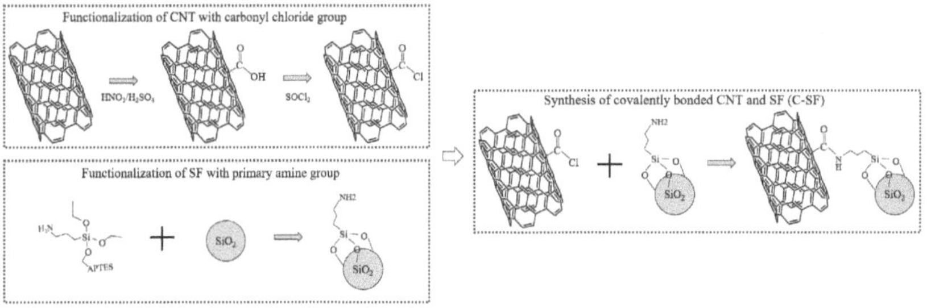

Fig. 5.6 Synthetic schemes of covalently synthesized CNT and silica fume [79].

have widely investigated the combined usage of several methods. For instance, Maria et al. [47] employed the ultrasonic method along with the surfactant method to study the dispersion of different length MWCNTs. They noted that the sonication method, when used alone, could not lead to proper dispersion while the combined use of sonication and surfactants resulted in much better dispersion of MWCNTs. They also investigated the effect of surfactant concentration on the dispersion of MWCNTs. They considered different surfactants to MWCNTs weight ratio of 0, 1.5, 4.0, and 6.25 in their study. As can be seen in Fig. 5.7, in spite of ultrasonication, CNT agglomeration can be observed when the surfactant was not used (Fig. 5.7A). However, by increasing the surfactant concentration, the dispersion of CNTs improved significantly (Fig. 5.7B–D).

As mentioned above, the synergistic dispersion effect of silica fume and polycarboxylic acid-based superplasticizer on CNTs was reported by Kim et al. [80]. They noted that adding superplasticizer reduced the van der Waals force among CNTs and increased the gaps, which consequently reduced agglomeration in cementitious composites (Fig. 5.8). Silica fume could further improve the dispersion of CNTs by filling the created gaps between CNTs. Researchers also have investigated the combination of the ultrasonic method, the surfactant method, and the covalent functionalizing method. Conducted research reveals that proper utilization of several methods can result in more effective dispersion of CNTs/CNFs and also decrease the dispersing time [81].

5.3.3 Novel routes of 1D nanomaterials dispersion

All the aforementioned dispersing methods for 1D nanomaterials suffer from some disadvantages, such as being time-consuming, consuming high energy,

Fig. 5.7 Surfactant concentration effect on nanotube dispersion: (A)–(D) represent a dispersant to MWCNT weight ratio of 0, 1.5, 4.0, and 6.25, respectively.

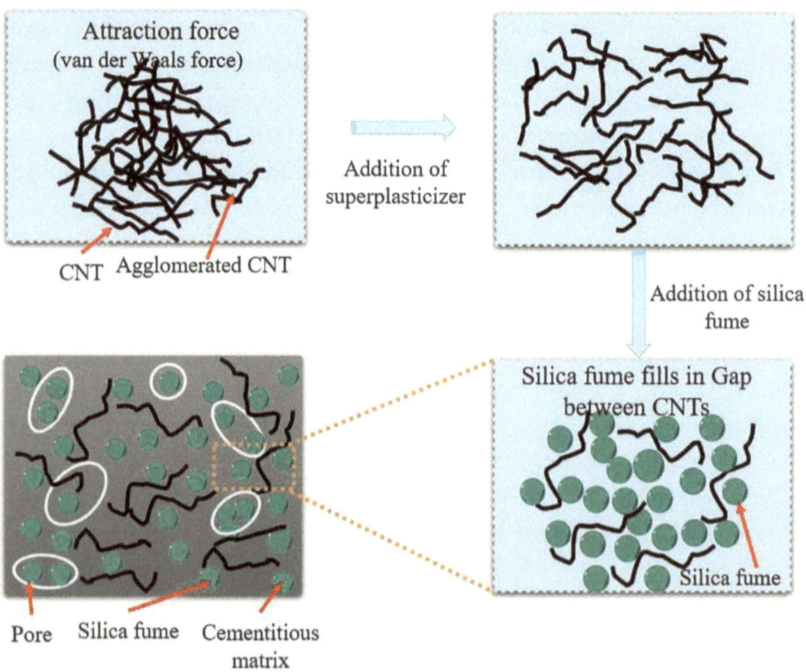

Fig. 5.8 Influences of superplasticizer and silica fume on the dispersion of CNT.

and negatively affecting mechanical properties of carbon nanomaterials [82]. Moreover, by utilizing conventional dispersing methods, it is not possible to add a high dosage of carbon nanomaterials into the cement matrix. Recently, researchers have proposed a novel method focusing on the in-situ growth of CNTs and CNFs on the cement and mineral admixture using chemical vapor deposition (CVD) and microwave irradiating conductive polymers method [83].

Chemical vapor deposition: CVD provides a relatively straightforward route for synthesizing 1D nanomaterials. CVD utilizes the decomposition or reaction of gas, liquid, or solid precursors under a controlled atmosphere to grow 1D nanomaterials [84,85]. For CVD, catalytically active substrates are generally most effective, and in this method, synthesis can be performed at pressures ranging from atmospheric to ultrahigh vacuum. Substrate interactions are vital to the growth of 1D nanomaterials [86,87] and can tune the properties of the material [88] through strain, symmetry, and chemical interactions.

The processes used in the CVD method to grow CNTs and CNFs on the cement and other mineral surfaces are as follows:

- a carbon source gas is passed through the reactor and decomposed
- adding proper catalyst
- a powder feeder is used to introduced cement and other mineral admixtures into the reactor
- the temperature should be maintained in the range of 400–700°C
- a powder collector should be used at the end of the reactor system to collect products

A schematic of the explained processes is shown in Fig. 5.9.

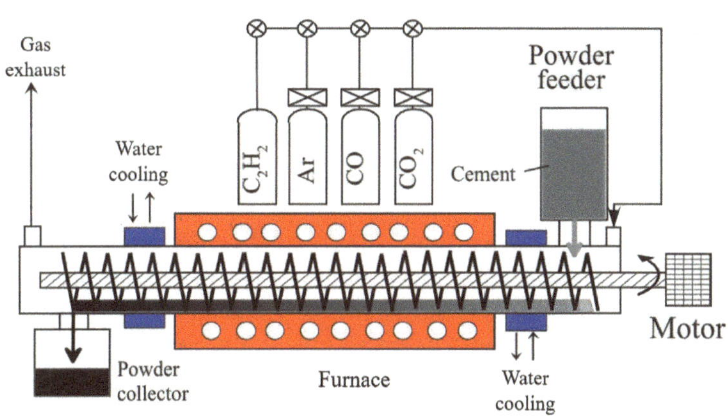

Fig. 5.9 Schematic view of the CVD method [89].

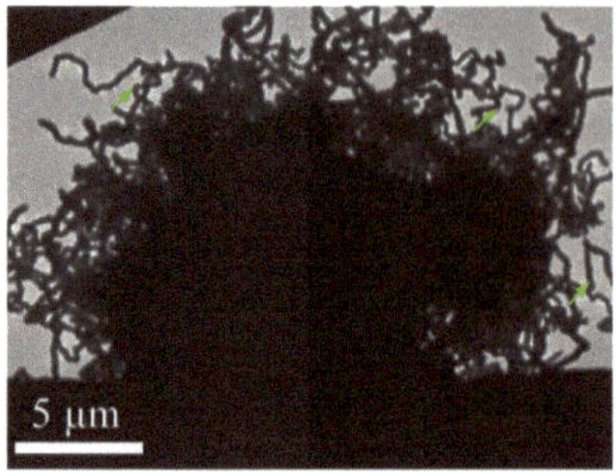

Fig. 5.10 The TEM image of complete coverage of the cement particles by CNTs/CNFs [89].

It has been shown that temperature plays an important role in the formation of CNTs and CNFs on the cement surface [89]. For instance, Nasibulin et al. [89–91] revealed that CNFs with the diameter of about 30 nm and average length of 3 μm could be grown on the surface of the cement particles (Fig. 5.10) in the temperature range from 550°C to 750°C, and that at the temperature below 450°C, no carbon grew on the cement particles. They also attempted to grow CNTs and CNFs on the surface of silica fume. They reported that the CNFs with an average diameter of 40 nm were observed at the temperature of 550°C, and that CNTs were observed at the temperature of 600°C (Fig. 5.11). The growth of CNTs and CNFs on the surface of cement clinker and silica fume by CVD method was studied by Ludvig et al. [92,93]. They revealed that high-purity CNTs could be obtained on the surface of cement clinker, as shown in Fig. 5.12. Additionally, CNTs and CNFs with a diameter of 30–60 nm were observed on the silica fume surface.

5.3.3.1 Microwave irradiating conductive polymer method

Another method that is currently employed in-situ growth of CNTs and CNFs is microwave irradiating conductive polymer method. In this method, a polymer material is used to absorb microwaves and is utilized as the heating source. Polypyrrole is the polymeric material that usually uses for the growth of nanocarbons due to its high efficiency in absorbing microwave irradiation. Powder precursors such as ferrocene are also mixed with polypyrrole in the solid state. The microwave is then radiated to the sample, which increases the temperature above 1100°C at which the ferrocene is

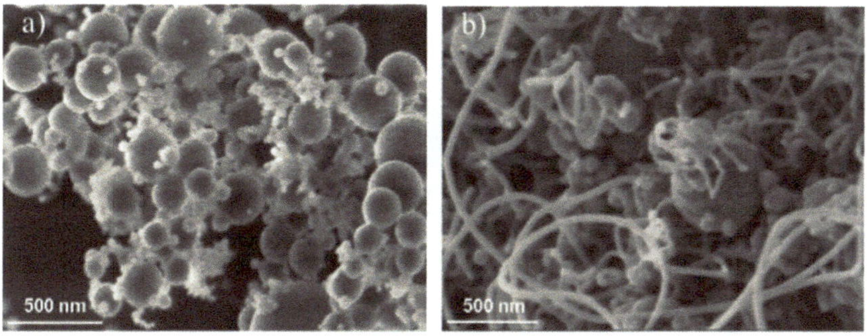

Fig. 5.11 SEM images of (A) pristine silica particles and (B) CNTs grown on the surface of silica [78].

Fig. 5.12 SEM image of CNTs grown on clinker [92].

decomposed to an iron catalyst and cyclopentadienyl. Cyclopentadienyl could also be used as a carbon source for initiating CNTs and CNFs growth (Fig. 5.13) [94–96].

Microwave irradiating conductive polymer technique utilizing for the in-situ growth of CNTs and CNFs was firstly used by Liu and Zhang [94–96]. They successfully grew the CNTs and CNFs on the surface of fly ash particles by using microwave irradiating conductive polymer method (Fig. 5.14).

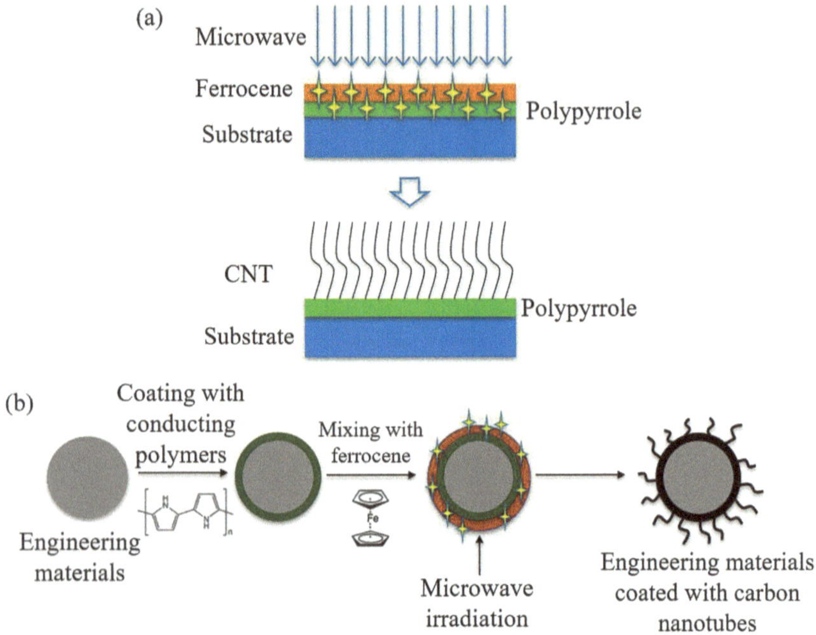

Fig. 5.13 Schematic view of the microwave irradiating conductive polymer method [94–96].

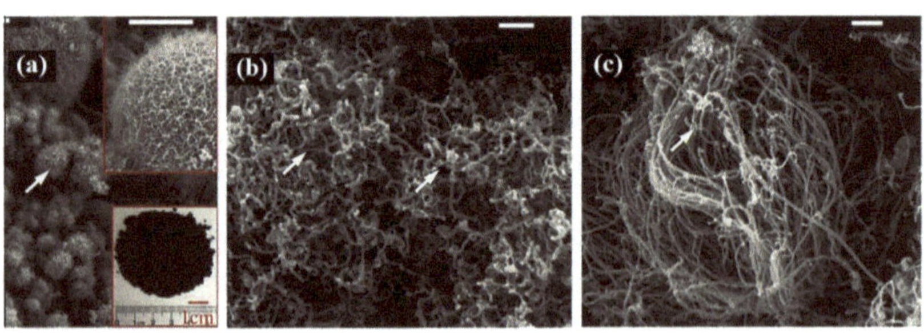

Fig. 5.14 (A) SEM images of the CNTs on fly ash, *(top)* zoom-in SEM image; *(bottom)* digital picture of coated fly ash (B) and (C) SEM images of CNTs coated polypyrrole coated fly ash [94–96].

5.4 Properties of cement-based materials with 1D nanomaterials

5.4.1 Rheological properties

The parameters affecting the rheological behavior of concrete have gained attention from researchers in recent years. Conducted research indicated that chemical composition, fineness of cement, geometrical parameters of solid components [97–99], properties of mineral and chemical admixtures [100–103], cement hydration, temperature, and humidity [104], as well as properties of filler [105,106] are some of the factors influencing the rheological characteristics of concrete.

Cepuritis et al. [106] used the sonication method to disperse MWCNTS (diameter 10–30 nm and length 1–2 μm) in the substitution level of 0.75% by cement weight and revealed that adding MWCNTS reduces the workability of the cement composite. Parveen et al. [75] used Pluronic and Defoamer as a surfactant and disperse MWCNTs (diameter < 8 nm and length 10–30 μm) in the substitution level of 0.1% by cement weight. They reported that workability decreased by 3.18%. Kang et al. [77] investigated the effect of adding MWCNTs (diameter 20 nm and length 1–25 μm), which functionalized with sulfuric acid (H_2SO_4) and nitric acid (HNO_3) in the substitution level of 0.1% by cement weight on the workability of cementitious mixtures. They dispersed MWCNTs with superplasticizer and sonication and revealed that the flow diameter of the mixture containing 0.1% MWCNTs reduced by 8.3% in comparison to the control mixtures.

In another study, Kim et al. [107] used MWCNTs (diameter 5–15 nm and length 10 μm) in the substitution level of 0.1%, 0.3%, and 0.5% by cement weight. Using a fixed dosage of superplasticizer, they dispersed MWCNTs in the mixtures with different water-to-binder ratio of 0.4, 0.5, and 0.6. Similar to the previous studies, they showed that adding MWCNTs decreased the workability of the mixtures at all water-to-binder ratios. They noted that the workability was further decreased by increasing the MWCNTs' content. In a similar study, Collins et al. [73] reported that by adding MWCNTs in the substitution level of 0.5%, 1%, and 2% by weight of cement, the slump diameter of the mixtures reduced by 14.5%, 32.8%, and 48.9%, respectively. Manzur et al. [108] investigated the effect of adding MWCNTs (diameter 10–20 nm and length 10–30 μm) on the rheological properties of cement composite. They disclosed that MWCNTs caused the setting time of the mixture to reduce by 25% in comparison to the control mixture. In addition, they revealed that adding MWCNTs could prevent bleeding in the mixture.

5.4.2 Mechanical properties

Ultra-high aspect ratio, along with excellent mechanical properties and chemical stability, has made CNTs and CNFs useful materials for reinforcing cement composite in comparison to traditional materials such as glass fiber and steel fiber. CNTs and CNFs improve the properties of cementitious composites through three primary mechanisms [109]:

- Crack-bridging mechanism
- Modifying the microstructure of hydration products and reducing the nanoporosity of cement paste
- Playing a role as nucleating agents for hydration gels

Researchers have widely investigated the enhancement in the strength of cement-based composites caused by adding CNFs and CNTs. Guan Fenglin [110] studied the rupture and compressive strength of mixtures containing different contents of CNFs at different ages. They showed that adding CNFs increased the 3 and 28 days compressive and rupture strength. Konsta-Gdoutos et al. [111] studied the tensile and compressive elastic modules of mortar containing different contents of CNTs and CNFs. They disclosed that in comparison to the control mixture, mortars incorporating CNTs and CNFs had higher tensile and compressive elastic modules.

The effects of CNTs (diameter 15–20 nm and length 0.1–10 μm) on the compressive and flexural properties of nonautoclaved and autoclaved aerated concretes were studied by Keriene et al. [112]. CNTs were dispersed using surfactant and sonication methods. Their findings showed 7.35%, 4.41%, 11.03%, and 8.08% increase in compressive strength of nonautoclaved aerated concretes containing 0.4%, 1%, 2%, and 4% CNTs, respectively. However, they revealed that in substitution level of 6%, compressive strength was decreased by 7.35%. A similar trend was also observed for flexural strength. For autoclaved aerated concretes, they disclosed that the compressive strength was increased by 3.57%, 10.7%, 14.7%, and 24.5% in the mixtures including 0.08%, 0.1%, 0.2%, and 2% CNTs, respectively. The flexural strength was shown to increase by 7.4%, 19.4%, 25%, and 24.1%, respectively. Zou et al. [113] studied Young's modulus, flexural strength, and fracture energy of cement composite containing functionalized CNTs (diameter 9.5 nm and length 1.5 μm) in the substitution level of 0.038% and 0.075%. Obtained results showed a significant improvement in Young's modulus, flexural strength, and fracture energy.

Wang et al. [114] investigated the flexural toughness index and fracture energy of cement composite incorporating CNTs (diameter 20–40 nm and length 5–15 μm). They used surfactant (anionic Arabic gum) and ultrasonic method to disperse the CNTs. The flexural toughness index was shown to increase by 31%, 57.47%, 47.13%, 31.61%, and 10.34% by incorporating 0.05%, 0.08%, 0.1%, 0.12%, and 0.15% CNTs, respectively. The fracture energy was also increased by 4.89%, 501.11%, 347%, 191.87%, and 34%, respectively. Shah et al. [44] investigated the effect aspect ratio of CNTs on the mechanical performance of concrete at ages 3, 7, and 28 days. They added 0.048% and 0.08% CNTs, by weight, to cement paste by using surfactants and sonication methods. Short CNTs with an aspect ratio of 700 and long CNTs with an aspect ratio of 1600 were used. Based on the reported results, to achieve effective reinforcement, lower amounts of CNTs (0.048%) were needed when using long CNTs, while higher amounts of CNTs (0.08%) were required when using short CNTs.

From the above-cited references, it can be concluded that adding CNTs and CNFs can improve the mechanical properties of cement materials, even in low concentration. In Table 5.1, the effects of adding CNTs and CNFs on the mechanical properties of cement composite are summarized.

5.4.3 Hydration kinetics and microstructure

As stated earlier, the effect of adding carbon nanomaterials on the mechanical properties of cementitious composite has been widely studied by researchers. Conducted research has revealed that carbon nanomaterials make significant changes to the microstructure of cement composite. Carbon nanomaterials affect the hydration kinetics and microstructure through different mechanisms that are discussed in the following paragraph.

Crack-bridging mechanism: It has been reported that CNTs and CNFs are able to bridge gaps and pores in the range of 10–1000 nm [75,127]. This bridging mechanism acts as the main reinforcing mechanism in cement matrices that makes delay to crack growth and hampers the formation of micro-size pores or gaps by transferring load from the cement matrix to the CNTs and CNFs [128]. The bridging effect, as shown in Fig. 5.15, creates a network structure between micro and nanocracks, which leads to an improvement in mechanical properties such as bending strength and elastic modulus [38].

Modifying the microstructure of hydration products and reducing the nanoporosity of cement paste: Another mechanism through which carbon nanomaterials affect the microstructure of cement composite is by affecting the hydration process and porosity of the cement matrix. For instance, it has been reported that CNTs can increase the hydration rate [129]. Stynoski et al. [130]

Table 5.1 Enhancement of CNTs to mechanical properties of cementitious materials.

Mechanical properties	Enhancement in properties (%)	Concentration (wt%)		Reference
		CNTs	CNFs	
Tensile strength	34.28	0.3	–	[92]
	19	0.5	–	[115]
	22–26	–	0.2	[116]
Compressive strength	19	0.5	–	[74]
	70	0.05	–	[117]
	11.03	0.02	–	[112]
	50	0.045–0.15	–	[42]
	29.5	0.2	–	[118]
	25	0.5	–	[73]
	22	0.1	–	[119]
	30	0.15	–	[120]
	10–20	0.5	–	[121]
	42.7	–	0.16	[122]
	300	–	0.4	[123]
Flexural strength	25	0.048–0.08		[74]
	35.4	0.2		[118]
	43.7	0.75		[124]
	34	0.5		[121]
	11.2	0.004		[112]
	82	–	0.2	[63]
	46	–	0.04	[55]
Young's modulus	227	0.048–0.08		[125]
	50	0.048–0.08		[44]
	68		1	[126]

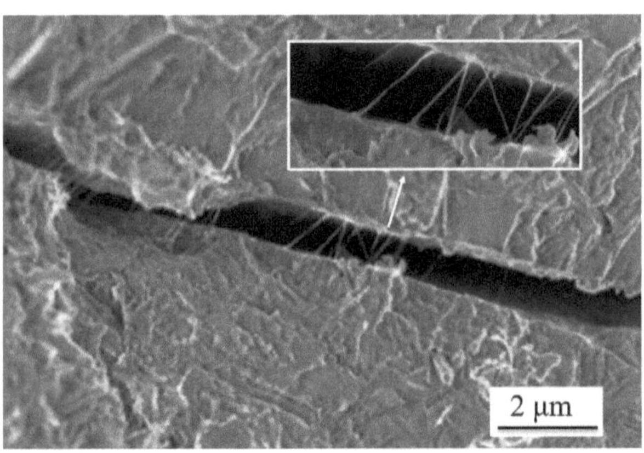

Fig. 5.15 CNTs act as crack-bridging [38].

showed that adding 0.05% MWCNTs (diameter 20–40 nm and length 0.5–40 μm), by volume, to cement caused the cement paste to reach its peak heat flow 3% sooner than the control mixture (Fig. 5.16).

Stynoski et al. [130] studied the effect of adding MWCNTs (diameter 20–50 nm and length 10–30 μm) in the substitution level of 0.25%, 0.5%, and 1% by weight, to cement pastes on the hydration process of cement composite. They disclosed that in the first 3 h, no obvious change to the hydration process of cement was observed. However, after 3 h, the first changes to the hydration process were observed, indicating that CNTs have increased the hydration rate. Peyvandi et al. [131] reported that (—COOH) covalence-modified CNFs formed a strong bond with Ca^{2+} ions in cement hydrates and improved interfacial interactions which consequently increased the mechanical properties of cement composite (Fig. 5.17).

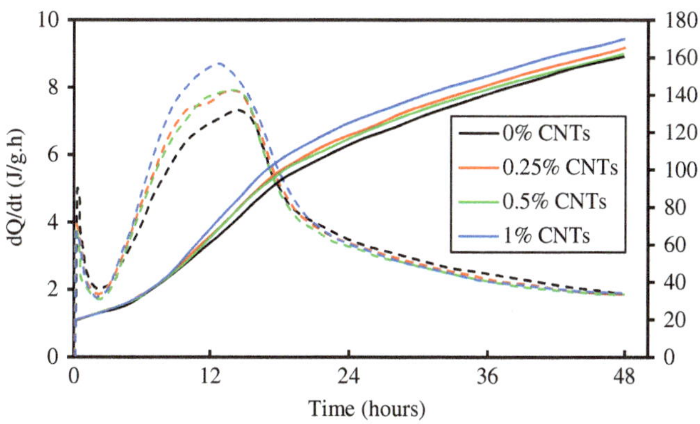

Fig. 5.16 The hydrothermal curve of cement-based composites with different CNTs dosage [130].

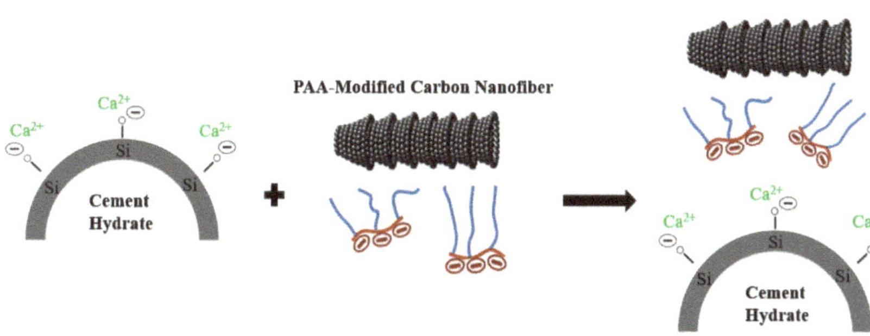

Fig. 5.17 Bonding between modified CNFs with cementitious material [131].

Additionally, CNTs and CNFs have shown to be capable of reducing porosity in the cement matrix by filling the gaps between the hydration products such as C—S—H gel and ettringite, which leads to the formation of denser microstructure [109]. Bharj et al. [119] showed that adding CNTs could improve the compressive and tensile strength by more than 30% by filling the pore in the range of 10–100 nm [77]. Manzur et al. [132] revealed that the size of CNTs is an influencing factor on the mechanical and microstructure properties of cement composite. They reported that more enhanced mechanical properties could be achieved by using finer scale CNTs due to higher efficiently of small-scale CNTs in filing pores in the cement matrix.

Playing a role as nucleating agents for hydration gels: It has been reported that CNTs and CNFs can act as a nucleating agent during the hardening process of cement [133]. The nucleation effect enhances the crystallinity of binding material, which in turn affects the flexural and compression strengths of cement composites [38]. Nucleation effect can improve the mechanical properties of cement composite through two mechanisms: 1—providing sites for the process of cement hydration which leads to generation of denser and stronger microstructure (Fig. 5.18), and 2—improving the bonding between CNTs and cement matrix by increasing the effective contact area between CNTs and the cement matrix [64].

5.4.4 Durability

The durability of cement-based materials is of great importance as it relates to the service life of buildings and structures. The porosity and pore characteristics of the cement matrix have been shown to have a significant influence on the durability of cement-based materials. In recent years, researchers have put considerable effort into using CNTs and CNFs to improve the durability of cement-based materials. It has been shown that CNTs and CNFs can improve the durability of cement-based materials by reducing the porosity and improving the pore structure of cement-based materials [38]. Additionally, CNTs and CNFs can improve the microstructure of the cement matrix, which, in turn, reduces the permeability of cement-based materials [38].

5.4.4.1 Freeze-thaw resistance

The effects of CNTs and CNFs on the freeze-thaw resistance of cement-based materials have been widely studied by researchers. Li et al. [135] studied the effect of adding carboxyl MWCNTs (diameter 10–20 nm and length

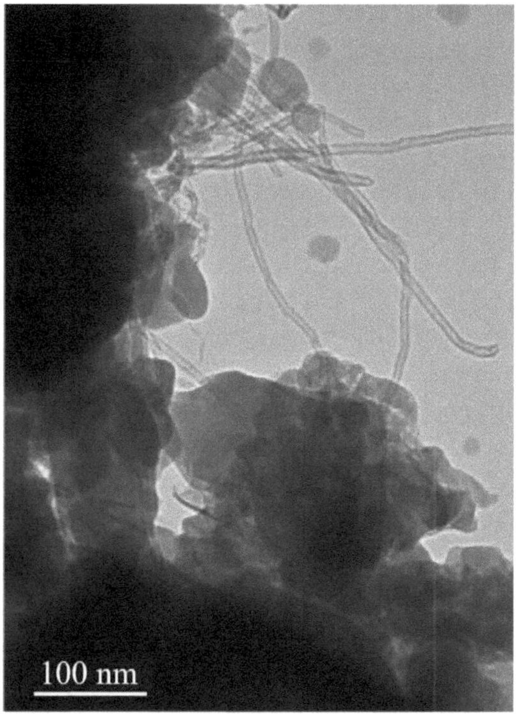

Fig. 5.18 TEM images of CNTs within the hardened cementitious composites [134].

10–30 μm) in the substitution level of 0.3% by weight, to mortars on the freeze–thaw resistance of cement composite. In their study, they exposed specimens to different cycles of freeze at −20°C and thaw in the water at 20°C at 28 days ages. They showed that after 30, 60, and 90 cycles of freeze-thaw, the specimen containing carboxyl MWCNTs had 26%, 40%, and 43% higher compressive strength than the control mixture. Cwirzen et al. [136] compared the effect of adding CNTs and CNFs in the substitution level of 0%–2.9% by weight, to cement on the freeze-thaw resistance of cement composite. CNTs were dispersed by using the surfactant and sonication method, while CNFs were directly added and mixed. CNFs were shown to be more effective than CNTs in improving the freeze-thaw resistance of mixtures.

5.4.4.2 Porosity, water absorption, and permeability

Hu et al. [137] investigated the effect of adding common CNTs and CNTs with a carboxyl group (CNTs-COOH) (diameter 10–20 nm and length

10–30 μm) in substitution level of 0.1% by weight of cement pastes on the porosity of cement composite. In this study, the combination of surfactant and sonication methods was employed to disperse CNTs. It was shown that the porosity was increased by 14.7% when using common CNTs while the porosity was decreased by 27.5% when using CNTs-COOH. Wang et al. [114] reported that using 0.12% MWCNTs (diameter 20–40 nm and length 5–15 μm) by weight to cement pastes led to approximately 3.25% reduction in the porosity. They used a surfactant-ultrasonic method with Arabic gum to disperse CNTs. Li et al. [135] revealed that utilizing carboxyl CNTs (diameter 10–20 nm and length 10–30 μm) in substitution level of 0.3%, by weight of cement, could refine pores with a diameter bigger than 200 nm to 50–100 nm and improve the disconnected pores.

Madhavi et al. [138] investigated the effect of adding CNTs (diameter 20–40 nm and length 1–10 μm), in substitution level of 0.015%, 0.03%, and 0.045%, by weight of cement, on the water absorption of cement composite. The surfactant and sonication methods were used to disperse CNTs. They disclosed that adding 0.015%, 0.03%, and 0.045% CNTs resulted in 10.22%, 14.41%, and 17.76% reduction of water absorption, respectively. Han et al. [139] reported that adding 0.2% carboxyl CNTs (diameter $<$ 8 nm and length 10–30 μm) by weight of cement led to a reduction in water absorption, water sorptivity, water permeability, and gas permeability of cement composite. They disclosed that this reduction was due to the modification of porosity.

5.4.4.3 Chloride ingress

Corrosion of rebars has been recognized as one of the most critical durability problems associated with reinforced concrete components that may be subjected to chloride environment. Accordingly, evaluating the permeability of concrete mixtures in a chloride environment can be highly beneficial. In this regard, Alafogianni et al. [140] investigated the effect of CNTs on the chloride ion permeation in the cement composite. They showed that CNTs did not have a significant effect on the resistance against the permeation of chloride ion in cement composite.

However, Wang et al. [141] comprehensively studied the effect of adding hydroxyl carbon nanotubes (CNTs-OH) and polyvinyl alcohol (PVA) on the corrosion of steel rebar embedded in the cement composite. CNTs-OH and PVA were used in the substitution level of 0.5% and 1% by weight of cement. They reported that utilizing CNTs-OH and PVA let to 29.9% and 83.2% reduction in the total porosity, respectively. Furthermore, as

Fig. 5.19 Distribution of rust in the splitting concrete (A) PCC, (B) CNT0.5, (C) PVA1.0, and (D) CNT0.5/PVA1.0) [142].

shown in Fig. 5.19, both CNTs-OH and PVA reduced the corrosion of embedded rebar (Fig. 5.19B and C). They showed that the integrated usage of CNTs-OH and PVA had the best performance and could reduce the corrosion level of embedded rebar by 94%. The reason for the higher efficiency of integrated using of CNTs-OH and PVA was reported to be the better dispersion of CNTs in PVA, pore-filling capability of both CNTs-OH and PVA, and restriction on the transportation of free ions.

Lee et al. [142] studied the combined effect of CNTs and nanosilica on the durability characteristic of cement mortars. In this study, 1% NS was used along with CNTs in the substation level of 0.01%–0.07% by weight of cement. They reported that the combined usage of NS and CNTs was effective in reducing the chloride ion penetration, as shown in Table 5.2.

Table 5.2 Rapid chloride ion penetration resistance of various mortars [143].

Systems	Charge passed (C)	Chloride ion penetration
Control	2577	Moderate
M1	1333	Low
M2	650	Very low
M3	457	Very low
M4	575	Very low
M5	679	Very low

M1 OPC + 1% NS-M2 OPC + 1% NS + 0.01% CNT-M3 OPC + 1% NS + 0.03% CNT-M4 OPC + 1% NS + 0.05% CNT-M5 OPC + 1% NS + 0.07% CNT.

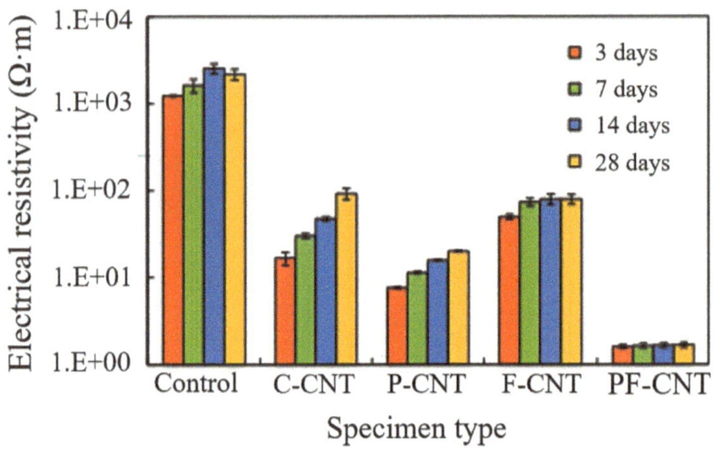

Fig. 5.20 Electrical resistivity of cementitious composites incorporating CNT at 3, 7, 14, and 28 days of curing [80].

5.4.5 Multifunctionality

5.4.5.1 Electrical and sensing properties

CNTs and CNFs have shown to have proper electrical conductivity. Therefore, using CNTs and CNFs in cement-based materials in order to enhance electrical conductivity has gained considerable attention from researchers.

Kim et al. [80] investigated the effect of silica fume and CNTs on the electrical properties of cement composite. In their study, they used a different combination of CNTs, silica fume, and superplasticizer (C-CNT: cement and CNTs, F-CNT: cement, CNTs and silica fume, P-CNT: cement, CNTs and superplasticizer, PF-CNT: cement, CNTs, superplasticizer and silica fume). As it is clear from Fig. 5.20, using CNTs resulted in a reduction in electrical resistivity of cement composite. Additionally, the combination of CNTS with silica fume showed the best performance in

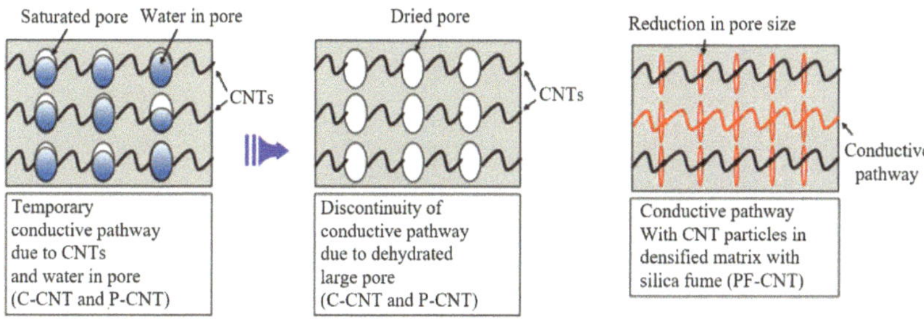

Fig. 5.21 The improved continuity of CNT particles by the addition of silica fume [80].

reducing electrical resistivity. They reported that utilizing SF could refine the pore size and fill the pores in the cement matrix. As shown in Fig. 5.21, silica fume replaced the entrapped air, and due to the higher electrical conductivity of SF than air, the overall electrical resistivity of the mixtures decreased.

Luo et al. [143] reported that the electrical resistivity of cement composite was decreased by 99.8% when using CNTs in the substitution level of 2% by weight of cement. Lou et al. [144] studied the variation in resistivity under the loaded stress (piezoresistive effect) in cement composite incorporating CNTs. They revealed that the piezoresistive effect provides cement-based materials with the benefit of being able to act as a sensor for perceiving the internal stress, strain, and damaged extent.

Cha et al. [145] reported that using functionalized CNTs could increase the piezoresistive sensitivity of cement by 150%. Li et al. [62] studied the effect of CNTs treated with acids (H_2SO_4 and HNO_3) and neat CNTs on the sensitivity of cement-based composites and reported that acid treatment could improve sensitivity (Fig. 5.22).

5.4.5.2 Thermal conductivity

Besides proper electrical conductivity, CNFs and CNTs possess proper thermal conductivity owing to their unique structural skeleton. Li et al. [146] reported that adding CNTs in the substitution level of 3%, by weight, to cement could significantly improve the thermal conductivity and heat-storage performance of cement-based materials. In a similar study, Han et al. [147] investigated the effect of CNTs, CNFs, and phase change materials (PCMs) on the thermal performance of cement-based materials. The obtained results indicated that better insulation performance could be achieved by using CNTs and CNFs.

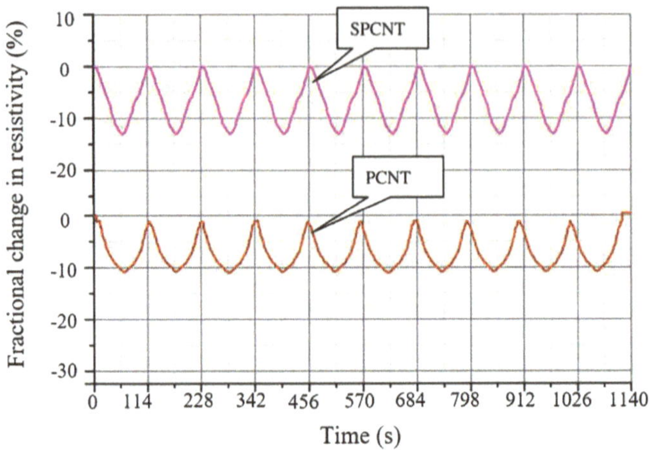

Fig. 5.22 The fractional change in resistivity versus time under cyclic compressive loading (0–15 kN) [62].

Xu et al. [148] revealed that adding CNTs to PCM-doped concrete had a slight effect on the thermal stability and chemical compatibility. However, they reported that CNTs significantly improved the thermal storage performance of PCM-doped concrete. Qin et al. [115] reported that incorporating high content of CNTs and CNFs enhanced the temperature sensitivity of cement-based materials.

5.4.5.3 Wave-absorbing performance

Singh et al. [57] studied the effect of adding CNTs (diameter 60 nm and length 15 μm) in different substitution levels to cement on the electromagnetic shielding properties of cement composite. The results indicated that in the substitution level of 15% shielding effectiveness values, more than 27 dB in X-band (8.2–12.4 GHz) could be achieved. Zhao et al. [149] disclosed that CNTs have the capability of wave-absorbing and revealed that the CNT's wave-absorbing performance improved by increasing the content of CNTs. Wang et al. [150] disclosed that CNTs have the ability to convert electromagnetic energy into heat energy and reported that improved wave-absorbing performance could be achieved by increasing CNTs' content. Micheli et al. [151] investigated the effect of adding CNTs (diameter

9.5 nm and length 1.5 μm) in the substitution level of 0.2% and 0.3% by weight to cement paste. The results indicated that adding CNTs could enhance shielding effectiveness and could attenuate the electromagnetic field. They showed that for a 15-cm thick wall containing 3% CNTs, shielding effectiveness values above 50 dB could be achieved.

5.5 Conclusions

The use of 1D nanomaterials in cementitious materials has been extensively studied in the laboratory in terms of their effect on fresh and hardened properties. The addition of 1D nanoparticles in paste, mortar, and concrete is a good way to enhance their properties and develop high-performance cementitious composites.

References

[1] S. Iijima, T. Ichihashi, Single-shell carbon nanotubes of 1-nm diameter, Nature 363 (6430) (1993) 603–605.
[2] B.M. Tyson, Carbon nanotube and Nanofiber Reinforcement for Improving the Flexural Strength and Fracture Toughness of Portland Cement Paste, Doctoral dissertation, Texas A & M University, 2012.
[3] S. Iijima, Helical microtubules of graphitic carbon, Nature 354 (6348) (1991) 56–58.
[4] J.P. Salvetat, J.M. Bonard, N.H. Thomson, A.J. Kulik, L. Forro, W. Benoit, L. Zuppiroli, Mechanical properties of carbon nanotubes, Appl. Phys. A 69 (3) (1999) 255–260.
[5] G. Dresselhaus, M.S. Dresselhaus, R. Saito, Physical properties of carbon nanotubes, World Sci. 1998 (1998).
[6] Q. Zhang, G. Chen, S.F. Yoon, J. Ahn, S.G. Wang, Q. Zhou, et al., Thermal conductivity of multiwalled carbon nanotubes, Phys. Rev. B 66 (16) (2002) 165440.
[7] S. Bal, S.S. Samal, Carbon nanotube reinforced polymer composites—a state of the art, Bull. Mater. Sci. 30 (4) (2007) 379–386.
[8] Z. Spitalsky, D. Tasis, K. Papagelis, C. Galiotis, Carbon nanotube–polymer composites: chemistry, processing, mechanical and electrical properties, Prog. Polym. Sci. 35 (3) (2010) 357–401.
[9] W.A. Curtin, B.W. Sheldon, CNT-reinforced ceramics and metals, Mater. Today 7 (11) (2004) 44–49.
[10] C. Liu, Y.Y. Fan, M. Liu, H.T. Cong, H.M. Cheng, M.S. Dresselhaus, Hydrogen storage in single-walled carbon nanotubes at room temperature, Science 286 (5442) (1999) 1127–1129.
[11] W.A. De Heer, A. Chatelain, D. Ugarte, A carbon nanotube field-emission electron source, Science 270 (5239) (1995) 1179–1180.
[12] L.H. Nguyen, T.V. Phi, P.Q. Phan, H.N. Vu, C. Nguyen-Duc, F. Fossard, Synthesis of multi-walled carbon nanotubes for NH_3 gas detection, Physica E 37 (1–2) (2007) 54–57.
[13] G.M. Kim, I.W. Nam, B. Yang, H.N. Yoon, H.K. Lee, S. Park, Carbon nanotube (CNT) incorporated cementitious composites for functional construction materials: the state of the art, Compos. Struct. 227 (2019) 111244.

[14] M.-F. Yu, B.S. Files, S. Arepalli, R.S. Ruoff, Tensile loading of ropes of single wall carbon nanotubes and their mechanical properties, Phys. Rev. Lett. 84 (24) (2000) 5552.
[15] V.G. Hadjiev, M.N. Iliev, S. Arepalli, P. Nikolaev, B.S. Files, Raman scattering test of single-wall carbon nanotube composites, Appl. Phys. Lett. 78 (21) (2001) 3193–3195.
[16] R. Siddique, A. Mehta, Effect of carbon nanotubes on properties of cement mortars, Constr. Build. Mater. 50 (2014) 116–129.
[17] T. Filleter, R. Bernal, S. Li, H.D. Espinosa, Ultrahigh strength and stiffness in cross-linked hierarchical carbon nanotube bundles, Adv. Mater. 23 (25) (2011) 2855–2860.
[18] B. Peng, M. Locascio, P. Zapol, S. Li, S.L. Mielke, G.C. Schatz, H.D. Espinosa, Measurements of near-ultimate strength for multiwalled carbon nanotubes and irradiation-induced crosslinking improvements, Nat. Nanotechnol. 3 (10) (2008) 626–631.
[19] J. Hone, Phonons and thermal properties of carbon nanotubes, Carbon Nanotubes (2001) 273–286.
[20] J. Maultzsch, S. Reich, C. Thomsen, E. Dobardžić, I. Milošević, M. Damnjanović, Phonon dispersion of carbon nanotubes, Solid State Commun. 121 (9–10) (2002) 471–474.
[21] M.A. Stroscio, M. Dutta, D. Kahn, K.W. Kim, Continuum model of optical phonons in a nanotube, Superlattice. Microst. 29 (6) (2001) 405–409.
[22] M. Grujicic, G. Cao, B. Gersten, Atomic-scale computations of the lattice contribution to thermal conductivity of single-walled carbon nanotubes, Mater. Sci. Eng. B 107 (2) (2004) 204–216.
[23] S.P. Hepplestone, A.M. Ciavarella, C. Janke, G.P. Srivastava, Size and temperature dependence of the specific heat capacity of carbon nanotubes, Surf. Sci. 600 (18) (2006) 3633–3636.
[24] Z.F. Ren, Z.P. Huang, J.W. Xu, J.H. Wang, P. Bush, M.P. Siegal, P.N. Provencio, Synthesis of large arrays of well-aligned carbon nanotubes on glass, Science 282 (5391) (1998) 1105–1107.
[25] T. Yamabe, K. Fukui, The Science and Technology of carbon Nanotubes, Elsevier, 1999.
[26] S.J. Tans, A.R. Verschueren, C. Dekker, Room-temperature transistor based on a single carbon nanotube, Nature 393 (6680) (1998) 49–52.
[27] B.Q. Wei, R. Vajtai, P.M. Ajayan, Reliability and current carrying capacity of carbon nanotubes, Appl. Phys. Lett. 79 (8) (2001) 1172–1174.
[28] H. Dai, E.W. Wong, C.M. Lieber, Probing electrical transport in nanomaterials: conductivity of individual carbon nanotubes, Science 272 (5261) (1996) 523–526.
[29] T.W. Ebbesen, H.J. Lezec, H. Hiura, J.W. Bennett, H.F. Ghaemi, T. Thio, Electrical conductivity of individual carbon nanotubes, Nature 382 (6586) (1996) 54–56.
[30] B.M. Tyson, Carbon nanotube and Nanofiber Reinforcement for Improving the Flexural Strength and Fracture Toughness of Portland Cement Paste, Doctoral dissertation, Texas A & M University, 2012.
[31] A.K. Geim, K.S. Novoselov, The rise of graphene, in: Nanoscience and Technology: A Collection of Reviews from Nature Journals, 2010, pp. 11–19.
[32] F. Sanchez, C. Ince, Microstructure and macroscopic properties of hybrid carbon nanofiber/silica fume cement composites, Compos. Sci. Technol. 69 (7-8) (2009) 1310–1318.
[33] E.T. Thostenson, C. Li, T.W. Chou, Nanocomposites in context, Compos. Sci. Technol. 65 (3–4) (2005) 491–516.
[34] T. Uchida, D.P. Anderson, M.L. Minus, S. Kumar, Morphology and modulus of vapor grown carbon nano fibers, J. Mater. Sci. 41 (18) (2006) 5851–5856.
[35] P.J. Goodhew, A.J. Clarke, J.E. Bailey, A review of the fabrication and properties of carbon fibres, Mater. Sci. Eng. 17 (1) (1975) 3–30.

[36] J. Shen, W. Huang, L. Wu, Y. Hu, M. Ye, Thermo-physical properties of epoxy nanocomposites reinforced with amino-functionalized multi-walled carbon nanotubes, Compos. A: Appl. Sci. Manuf. 38 (5) (2007) 1331–1336.
[37] W. Meng, K.H. Khayat, Mechanical properties of ultra-high-performance concrete enhanced with graphite nanoplatelets and carbon nanofibers, Compos. Part B: Eng. 107 (2016) 113–122.
[38] T. Shi, Z. Li, J. Guo, H. Gong, C. Gu, Research progress on CNTs/CNFs-modified cement-based composites–a review, Constr. Build. Mater. 202 (2019) 290–307.
[39] S. Parveen, S. Rana, R. Fangueiro, A review on nanomaterial dispersion, microstructure, and mechanical properties of carbon nanotube and nanofiber reinforced cementitious composites, J. Nanomater. 2013 (2013).
[40] W. Meng, K.H. Khayat, Effect of graphite nanoplatelets and carbon nanofibers on rheology, hydration, shrinkage, mechanical properties, and microstructure of UHPC, Cem. Concr. Res. 105 (2018) 64–71.
[41] S. Parveen, S. Rana, R. Fangueiro, A review on nanomaterial dispersion, microstructure, and mechanical properties of carbon nanotube and nanofiber reinforced cementitious composites, J. Nanomater. 2013 (2013).
[42] A. Cwirzen, K. Habermehl-Cwirzen, V. Penttala, Surface decoration of carbon nanotubes and mechanical properties of cement/carbon nanotube composites, Adv. Cem. Res. 20 (2) (2008) 65–73.
[43] S.P. Shah, M.S. Konsta-Gdoutos, Z.S. Metaxa, Advanced cement based nanocomposites, in: Recent Advances in Mechanics, Springer, Dordrecht, 2011, pp. 313–327.
[44] S.P. Shah, M.S. Konsta-Gdoutos, Z.S. Metaxa, P. Mondal, Nanoscale modification of cementitious materials, in: Nanotechnology in Construction 3, Springer, Berlin, Heidelberg, 2009, pp. 125–130.
[45] F. Azhari, N. Banthia, Cement-based sensors with carbon fibers and carbon nanotubes for piezoresistive sensing, Cem. Concr. Compos. 34 (7) (2012) 866–873.
[46] J. Keriene, M. Kligys, A. Laukaitis, G. Yakovlev, A. Špokauskas, M. Aleknevičius, The influence of multi-walled carbon nanotubes additive on properties of non-autoclaved and autoclaved aerated concretes, Constr. Build. Mater. 49 (2013) 527–535.
[47] M.S. Konsta-Gdoutos, Z.S. Metaxa, S.P. Shah, Highly dispersed carbon nanotube reinforced cement based materials, Cem. Concr. Res. 40 (7) (2010) 1052–1059.
[48] Z.S. Metaxa, J.W.T. Seo, M.S. Konsta-Gdoutos, M.C. Hersam, S.P. Shah, Highly concentrated carbon nanotube admixture for nano-fiber reinforced cementitious materials, Cem. Concr. Compos. 34 (5) (2012) 612–617.
[49] M.S. Konsta-Gdoutos, Z.S. Metaxa, S.P. Shah, Multi-scale mechanical and fracture characteristics and early-age strain capacity of high performance carbon nanotube/cement nanocomposites, Cem. Concr. Compos. 32 (2) (2010) 110–115.
[50] O. Mendoza, G. Sierra, J.I. Tobón, Influence of super plasticizer and Ca(OH)2 on the stability of functionalized multi-walled carbon nanotubes dispersions for cement composites applications, Constr. Build. Mater. 47 (2013) 771–778.
[51] J. Yu, N. Grossiord, C.E. Koning, J. Loos, Controlling the dispersion of multi-wall carbon nanotubes in aqueous surfactant solution, Carbon 45 (3) (2007) 618–623.
[52] J. Luo, Z. Duan, H. Li, The influence of surfactants on the processing of multi-walled carbon nanotubes in reinforced cement matrix composites, Phys. Status Solidi A 206 (12) (2009) 2783–2790.
[53] Z.S. Metaxa, M.S. Konsta-Gdoutos, S.P. Shah, Carbon nanofiber cementitious composites: effect of debulking procedure on dispersion and reinforcing efficiency, Cem. Concr. Compos. 36 (2013) 25–32.
[54] A. Yazdanbakhsh, Z. Grasley, B. Tyson, R. Abu Al-Rub, Challenges and benefits of utilizing carbon nanofilaments in cementitious materials, J. Nanomater. 2012 (2012).

[55] A. Peyvandi, L.A. Sbia, P. Soroushian, K. Sobolev, Effect of the cementitious paste density on the performance efficiency of carbon nanofiber in concrete nanocomposite, Constr. Build. Mater. 48 (2013) 265–269.
[56] A.P. Singh, M. Mishra, A. Chandra, S.K. Dhawan, Graphene oxide/ferrofluid/cement composites for electromagnetic interference shielding application, Nanotechnology 22 (46) (2011) 465701.
[57] A.P. Singh, B.K. Gupta, M. Mishra, A. Chandra, R.B. Mathur, S.K. Dhawan, Multiwalled carbon nanotube/cement composites with exceptional electromagnetic interference shielding properties, Carbon 56 (2013) 86–96.
[58] A.L. Materazzi, F. Ubertini, A. D'Alessandro, Carbon nanotube cement-based transducers for dynamic sensing of strain, Cem. Concr. Compos. 37 (2013) 2–11.
[59] F. Ubertini, A.L. Materazzi, A. D'Alessandro, S. Laflamme, Natural frequencies identification of a reinforced concrete beam using carbon nanotube cement-based sensors, Eng. Struct. 60 (2014) 265–275.
[60] T. Hielscher, Ultrasonic production of nano-size dispersions and emulsions, arXiv preprint arXiv:0708.1831 (2007).
[61] P.C. Ma, N.A. Siddiqui, G. Marom, J.K. Kim, Dispersion and functionalization of carbon nanotubes for polymer-based nanocomposites: a review, Compos. A: Appl. Sci. Manuf. 41 (10) (2010) 1345–1367.
[62] G.Y. Li, P.M. Wang, X. Zhao, Pressure-sensitive properties and microstructure of carbon nanotube reinforced cement composites, Cem. Concr. Compos. 29 (5) (2007) 377–382.
[63] B.M. Tyson, R.K. Abu Al-Rub, A. Yazdanbakhsh, Z. Grasley, Carbon nanotubes and carbon nanofibers for enhancing the mechanical properties of nanocomposite cementitious materials, J. Mater. Civ. Eng. 23 (7) (2011) 1028–1035.
[64] K.M. Liew, M.F. Kai, L.W. Zhang, Carbon nanotube reinforced cementitious composites: an overview, Compos. A: Appl. Sci. Manuf. 91 (2016) 301–323.
[65] Y.H. Liao, O. Marietta-Tondin, Z. Liang, C. Zhang, B. Wang, Investigation of the dispersion process of SWNTs/SC-15 epoxy resin nanocomposites, Mater. Sci. Eng. A 385 (1–2) (2004) 175–181.
[66] S.M. Yuen, C.C.M. Ma, Y.Y. Lin, H.C. Kuan, Preparation, morphology and properties of acid and amine modified multiwalled carbon nanotube/polyimide composite, Compos. Sci. Technol. 67 (11−12) (2007) 2564–2573.
[67] M.T. Byrne, W.P. McNamee, Y.K. Gun'ko, Chemical functionalization of carbon nanotubes for the mechanical reinforcement of polystyrene composites, Nanotechnology 19 (41) (2008) 415707.
[68] K.L. Lu, R.M. Lago, Y.K. Chen, M.L.H. Green, P.J.F. Harris, S.C. Tsang, Mechanical damage of carbon nanotubes by ultrasound, Carbon 34 (6) (1996).
[69] K. Mukhopadhyay, C.D. Dwivedi, G.N. Mathur, Conversion of carbon nanotubes to carbon nanofibers by sonication, Carbon 8 (40) (2002) 1373–1376.
[70] A. Sobolkina, V. Mechtcherine, V. Khavrus, D. Maier, M. Mende, M. Ritschel, A. Leonhardt, Dispersion of carbon nanotubes and its influence on the mechanical properties of the cement matrix, Cem. Concr. Compos. 34 (10) (2012) 1104–1113.
[71] R. Rastogi, R. Kaushal, S.K. Tripathi, A.L. Sharma, I. Kaur, L.M. Bharadwaj, Comparative study of carbon nanotube dispersion using surfactants, J. Colloid Interface Sci. 328 (2) (2008) 421–428.
[72] A. Yazdanbakhsh, Z.C. Grasley, B. Tyson, R.A. Al-Rub, Carbon nano filaments in cementitious materials: some issues on dispersion and interfacial bond, Spec. Pub. 267 (2009) 21–34.
[73] F. Collins, J. Lambert, W.H. Duan, The influences of admixtures on the dispersion, workability, and strength of carbon nanotube–OPC paste mixtures, Cem. Concr. Compos. 34 (2) (2012) 201–207.

[74] G.Y. Li, P.M. Wang, X. Zhao, Mechanical behavior and microstructure of cement composites incorporating surface-treated multi-walled carbon nanotubes, Carbon 43 (6) (2005) 1239–1245.
[75] S. Parveen, S. Rana, R. Fangueiro, M.C. Paiva, Microstructure and mechanical properties of carbon nanotube reinforced cementitious composites developed using a novel dispersion technique, Cem. Concr. Res. 73 (2015) 215–227.
[76] G.M. Kim, I.W. Nam, H.N. Yoon, H.K. Lee, Effect of superplasticizer type and siliceous materials on the dispersion of carbon nanotube in cementitious composites, Compos. Struct. 185 (2018) 264–272.
[77] S.T. Kang, J.Y. Seo, S.H. Park, The characteristics of CNT/cement composites with acid-treated MWCNTs, Adv. Mater. Sci. Eng. 2015 (2015).
[78] P.R. Mudimela, L.I. Nasibulina, A.G. Nasibulin, A. Cwirzen, M. Valkeapää, K. Habermehl-Cwirzen, et al., Synthesis of carbon nanotubes and nanofibers on silica and cement matrix materials, J. Nanomater. 2009 (2009) 29.
[79] G.M. Kim, Y.K. Kim, Y.J. Kim, J.H. Seo, B.J. Yang, H.K. Lee, Enhancement of the modulus of compression of calcium silicate hydrates via covalent synthesis of CNT and silica fume, Constr. Build. Mater. 198 (2019) 218–225.
[80] G.M. Kim, B.J. Yang, K.J. Cho, E.M. Kim, H.K. Lee, Influences of CNT dispersion and pore characteristics on the electrical performance of cementitious composites, Compos. Struct. 164 (2017) 32–42.
[81] X. Wu, J. Xiao, F. Xia, Y. Hu, Z. Peng, Dispersion methods and dispersion mechanism of carbon nanotubes, Mater. Rev. 9 (2011) 16–19.
[82] W. Zhu, P.J. Bartos, A. Porro, Application of nanotechnology in construction, Mater. Struct. 37 (9) (2004) 649–658.
[83] S. Sun, X. Yu, B. Han, J. Ou, In situ growth of carbon nanotubes/carbon nanofibers on cement/mineral admixture particles: a review, Constr. Build. Mater. 49 (2013) 835–840.
[84] X. Li, W. Cai, J. An, S. Kim, J. Nah, D. Yang, et al., Large-area synthesis of high-quality and uniform graphene films on copper foils, Science 324 (5932) (2009) 1312–1314.
[85] J.H. Lee, E.K. Lee, W.J. Joo, Y. Jang, B.S. Kim, J.Y. Lim, et al., Wafer-scale growth of single-crystal monolayer graphene on reusable hydrogen-terminated germanium, Science 344 (6181) (2014) 286–289.
[86] R.S. Edwards, K.S. Coleman, Graphene film growth on polycrystalline metals, Acc. Chem. Res. 46 (1) (2012) 23–30.
[87] A. Koma, Van der Waals epitaxy—a new epitaxial growth method for a highly lattice-mismatched system, Thin Solid Films 216 (1) (1992) 72–76.
[88] C.R. Dean, et al., Boron nitride substrates for high-quality graphene electronics, Nat. Nanotechnol. 5 (10) (2010) 722.
[89] A.G. Nasibulin, S.D. Shandakov, L.I. Nasibulina, A. Cwirzen, P.R. Mudimela, K. Habermehl-Cwirzen, et al., A novel cement-based hybrid material, New J. Phys. 11 (2) (2009), 023013.
[90] A.G. Nasibulin, P.V. Pikhitsa, H. Jiang, D.P. Brown, A.V. Krasheninnikov, A.S. Anisimov, et al., A novel hybrid carbon material, Nat. Nanotechnol. 2 (3) (2007) 156.
[91] A.G. Nasibulin, T. Koltsova, L.I. Nasibulina, I.V. Anoshkin, A. Semencha, O.V. Tolochko, E.I. Kauppinen, A novel approach to composite preparation by direct synthesis of carbon nanomaterial on matrix or filler particles, Acta Mater. 61 (6) (2013) 1862–1871.
[92] P. Ludvig, L.O. Ladeira, J.M. Calixto, I.C.P. Gaspar, V.S. Melo, In-situ synthesis of multiwall carbon nanotubes on Portland cement clinker, in: 11th International Conference on Advanced Materials, Rio de Janeiro, Brazil, 2009.
[93] P. Ludvig, J.M. Calixto, L.O. Ladeira, I.C. Gaspar, Using converter dust to produce low cost cementitious composites by in situ carbon nanotube and nanofiber synthesis, Materials 4 (3) (2011) 575–584.

[94] Z. Liu, J. Wang, V. Kushvaha, S. Poyraz, H. Tippur, S. Park, et al., Poptube approach for ultrafast carbon nanotube growth, Chem. Commun. 47 (35) (2011) 9912–9914.
[95] X. Zhang, Z. Liu, Recent advances in microwave initiated synthesis of nanocarbon materials, Nanoscale 4 (3) (2012) 707–714.
[96] X. Zhang, S.K. Manohar, Bulk synthesis of polypyrrole nanofibers by a seeding approach, J. Am. Chem. Soc. 126 (40) (2004) 12714–12715.
[97] H. Vikan, H. Justnes, F. Winnefeld, R. Figi, Correlating cement characteristics with rheology of paste, Cem. Concr. Res. 37 (11) (2007) 1502–1511.
[98] J.J. Chen, A.K.H. Kwan, Superfine cement for improving packing density, rheology and strength of cement paste, Cem. Concr. Compos. 34 (1) (2012) 1–10.
[99] D. Nagrockiene, I. Pundienė, A. Kicaite, The effect of cement type and plasticizer addition on concrete properties, Constr. Build. Mater. 45 (2013) 324–331.
[100] C.F. Ferraris, K.H. Obla, R. Hill, The influence of mineral admixtures on the rheology of cement paste and concrete, Cem. Concr. Res. 31 (2) (2001) 245–255.
[101] N. Mikanovic, C. Jolicoeur, Influence of superplasticizers on the rheology and stability of limestone and cement pastes, Cem. Concr. Res. 38 (7) (2008) 907–919.
[102] C. Brumaud, R. Baumann, M. Schmitz, M. Radler, N. Roussel, Cellulose ethers and yield stress of cement pastes, Cem. Concr. Res. 55 (2014) 14–21.
[103] H. Bessaies-Bey, R. Baumann, M. Schmitz, M. Radler, N. Roussel, Effect of polyacrylamide on rheology of fresh cement pastes, Cem. Concr. Res. 76 (2015) 98–106.
[104] W.U. Di, S.J. Cai, G. Huang, Coupled effect of cement hydration and temperature on rheological properties of fresh cemented tailings backfill slurry, Trans. Nonferrous Metals Soc. China 24 (9) (2014) 2954–2963.
[105] R. Talero, C. Pedrajas, M. González, C. Aramburo, A. Blázquez, V. Rahhal, Role of the filler on Portland cement hydration at very early ages: rheological behaviour of their fresh cement pastes, Constr. Build. Mater. 151 (2017) 939–949.
[106] R. Cepuritis, S. Jacobsen, B. Pedersen, E. Mørtsell, Crushed sand in concrete–effect of particle shape in different fractions and filler properties on rheology, Cem. Concr. Compos. 71 (2016) 26–41. A. Hunashyal, N. Banapurmath, A. Jain, S. Quadri, A. Shettar, Experimental investigation on the effect of multi-walled carbon nanotubes and nano-SiO_2 addition on mechanical properties of hardened cement paste, Adv. Mater. 3 (5) (2014) 45–51.
[107] H.K. Kim, I.S. Park, H.K. Lee, Improved piezoresistive sensitivity and stability of CNT/cement mortar composites with low water–binder ratio, Compos. Struct. 116 (2014) 713–719.
[108] T. Manzur, N. Yazdani, M.A.B. Emon, Potential of carbon nanotube reinforced cement composites as concrete repair material, J. Nanomater. 2016 (2016) 2.
[109] B. Han, S. Sun, S. Ding, L. Zhang, X. Yu, J. Ou, Review of nanocarbon-engineered multifunctional cementitious composites, Compos. A: Appl. Sci. Manuf. 70 (2015) 69–81.
[110] L. Guan, J.X. Wu, Y. Gao, C.Y. Han, Effect of carbon nanofibers on mechanical and electrical properties of mortar, Low Temp. Archit. Technol. 09 (2016) 11–13.
[111] M.S. Konsta-Gdoutos, G. Batis, P.A. Danoglidis, A.K. Zacharopoulou, E.K. Zacharopoulou, M.G. Falara, S.P. Shah, Effect of CNT and CNF loading and count on the corrosion resistance, conductivity and mechanical properties of nanomodified OPC mortars, Constr. Build. Mater. 147 (2017) 48–57.
[112] J. Kerienė, M. Kligys, A. Laukaitis, G. Yakovlev, A. Špokauskas, M. Aleknevičius, The influence of multi-walled carbon nanotubes additive on properties of non-autoclaved and autoclaved aerated concretes, Constr. Build. Mater. 49 (2013) 527–535.
[113] B. Zou, S.J. Chen, A.H. Korayem, F. Collins, C.M. Wang, W.H. Duan, Effect of ultrasonication energy on engineering properties of carbon nanotube reinforced cement pastes, Carbon 85 (2015) 212–220.
[114] B. Wang, Y. Han, S. Liu, Effect of highly dispersed carbon nanotubes on the flexural toughness of cement-based composites, Constr. Build. Mater. 46 (2013) 8–12.

[115] A.M. Hunashyal, S.V. Tippa, S.S. Quadri, N.R. Banapurmath, Experimental investigation on effect of carbon nanotubes and carbon fibres on the behavior of plain cement mortar composite round bars under direct tension, ISRN Nanotechnol. 2011 (2011).
[116] C. Gay, F. Sanchez, Performance of carbon nanofiber–cement composites with a high-range water reducer, Transp. Res. Rec. 2142 (1) (2010) 109–113.
[117] G. Yakovlev, J. Keriené, A. Gailius, I. Girnienė, Cement based foam concrete reinforced by carbon nanotubes, Mater. Sci. 12 (2) (2006) 147–151.
[118] J.L. Luo, Fabrication and Functional Properties of Multi-Walled Carbon Nanotube/Cement Composites, Harbin Institute of Technology, Harbin, China, 2009. Dissertation for the Doctoral Degree in Engineering.
[119] J. Bharj, Experimental study on compressive strength of cement-CNT composite paste, Indian J. Pure Appl. Phys. 52 (1) (2015) 35–38.
[120] H.K. Kim, I.W. Nam, H.K. Lee, Enhanced effect of carbon nanotube on mechanical and electrical properties of cement composites by incorporation of silica fume, Compos. Struct. 107 (2014) 60–69.
[121] S. Musso, J.M. Tulliani, G. Ferro, A. Tagliaferro, Influence of carbon nanotubes structure on the mechanical behavior of cement composites, Compos. Sci. Technol. 69 (11–12) (2009) 1985–1990.
[122] D. Gao, M. Sturm, Y.L. Mo, Electrical resistance of carbon-nanofiber concrete, Smart Mater. Struct. 18 (9) (2009), 095039.
[123] L.I. Nasibulina, I.V. Anoshkin, A.V. Semencha, O.V. Tolochko, J.E. Malm, M.J. Karppinen, et al., Carbon nanofiber/clinker hybrid material as a highly efficient modificator of mortar mechanical properties, Mater. Phys. Mech. 13 (1) (2012) 77–84.
[124] A.M. Hunashyal, G.V. Sundeep, S.S. Quadri, N.R. Banapurmath, Experimental investigations to study the effect of carbon nanotubes reinforced in cement-based matrix composite beams, Proc. Inst. Mech. Eng., Part N 225 (1) (2011) 17–22.
[125] Y. Saez de Ibarra, J.J. Gaitero, E. Erkizia, I. Campillo, Atomic force microscopy and nanoindentation of cement pastes with nanotube dispersions, Phys. Status Solidi A 203 (6) (2006) 1076–1081.
[126] A. Yazdanbakhsh, Production, Characterization, and Mechanical Behavior of Cementitious Materials Incorporating carbon Nanofibers, Dissertation for the Doctoral Degree in Civil Engineering, Texas A&M University, Texas, America, 2013.
[127] Á. Kukovecz, G. Kozma, Z. Kónya, Multi-walled carbon nanotubes, in: Springer Handbook of Nanomaterials, Springer, Berlin, Heidelberg, 2013, pp. 147–188.
[128] S.C. Chowdhury, B.G. Haque, T. Okabe, J.W. Gillespie Jr., Modeling the effect of statistical variations in length and diameter of randomly oriented CNTs on the properties of CNT reinforced nanocomposites, Compos. Part B 43 (4) (2012) 1756–1762.
[129] J. Makar, The effect of SWCNT and other nanomaterials on cement hydration and reinforcement, in: Nanotechnology in Civil Infrastructure, Springer, Berlin, Heidelberg, 2011, pp. 103–130.
[130] P. Stynoski, P. Mondal, C. Marsh, Effects of silica additives on fracture properties of carbon nanotube and carbon fiber reinforced Portland cement mortar, Cem. Concr. Compos. 55 (2015) 232–240. H. Cui, S. Yang, S. Memon, Development of carbon nanotube modified cement paste with microencapsulated phase-change material for structural–functional integrated application, Int. J. Mol. Sci. 16 (4) (2015) 8027–8039.
[131] A. Peyvandi, P. Soroushian, N. Abdol, A.M. Balachandra, Surface-modified graphite nanomaterials for improved reinforcement efficiency in cementitious paste, Carbon 63 (2013) 175–186.
[132] T. Manzur, N. Yazdani, M. Emon, A. Bashar, Effect of carbon nanotube size on compressive strengths of nanotube reinforced cementitious composites, J. Mater. 2014 (2014).
[133] J.M. Makar, G.W. Chan, Growth of cement hydration products on single-walled carbon nanotubes, J. Am. Ceram. Soc. 92 (6) (2009) 1303–1310.
[134] R.K.A. Al-Rub, A.I. Ashour, B.M. Tyson, On the aspect ratio effect of multi-walled carbon nanotube reinforcements on the mechanical properties of cementitious nanocomposites, Constr. Build. Mater. 35 (2012) 647–655.

[135] W.W. Li, W.M. Ji, Y.C. Wang, Y. Liu, R.X. Shen, F. Xing, Investigation on the mechanical properties of a cement-based material containing carbon nanotube under drying and freeze-thaw conditions, Materials 8 (12) (2015) 8780–8792.

[136] A. Cwirzen, K. Habermehl-Cwirzen, The effect of carbon nano-and microfibers on strength and residual cumulative strain of mortars subjected to freeze-thaw cycles, J. Adv. Concr. Technol. 11 (3) (2013) 80–88.

[137] Y. Hu, D. Luo, P. Li, Q. Li, G. Sun, Fracture toughness enhancement of cement paste with multi-walled carbon nanotubes, Constr. Build. Mater. 70 (2014) 332–338.

[138] T.C. Madhavi, P. Pavithra, S.B. Singh, S.B. Vamsi Raj, S. Paul, Effect of multiwalled carbon nanotubes on mechanical properties of concrete, Int. J. Sci. Res. 2 (6) (2013) 166–168.

[139] B. Han, K. Zhang, X. Yu, E. Kwon, J. Ou, Fabrication of piezoresistive CNT/CNF cementitious composites with superplasticizer as dispersant, J. Mater. Civ. Eng. 24 (6) (2011) 658–665.

[140] P. Alafogianni, P.T. Dalla, I.K. Tragazikis, N.M. Barkoula, T.E. Matikas, Rapid chloride permeability test for durability study of carbon nanoreinforced mortar, in: Smart Sensor Phenomena, Technology, Networks, and Systems Integration 2015, vol. 9436, International Society for Optics and Photonics, 2015, p. 94360I.

[141] Z. Wang, J. Yu, G. Li, M. Zhang, C.K. Leung, Corrosion behavior of steel rebar embedded in hybrid CNTs-OH/polyvinyl alcohol modified concrete under accelerated chloride attack, Cem. Concr. Compos. 100 (2019) 120–129.

[142] H.S. Lee, B. Balasubramanian, G.V.T. Gopalakrishna, S.J. Kwon, S.P. Karthick, V. Saraswathy, Durability performance of CNT and nanosilica admixed cement mortar, Constr. Build. Mater. 159 (2018) 463–472.

[143] J.L. Luo, Z.D. Duan, T.J. Zhao, Properties of electrical resistivity of fiber reinforced cement composites with multi-walled carbon nanotubes, Harbin Gongye Daxue Xuebao/J. Harbin Instit. Technol. 42 (8) (2010) 1237–1241.

[144] J.L. Luo, Z.D. Duan, T.J. Zhao, Q.Y. Li, Effect of compressive strain on electrical resistivity of carbon nanotube cement-based composites, in: Key Engineering Materials, vol. 483, Trans Tech Publications, 2011, pp. 579–583.

[145] S.W. Cha, C. Song, Y.H. Cho, S. Choi, Piezoresistive Properties of CNT Reinforced Cementitious Composites, Maney Publisher, 2014, pp. S2716–S2721.

[146] H. Li, Q. Zhang, H. Xiao, Self-deicing road system with a CNFP high-efficiency thermal source and MWCNT/cement-based high-thermal conductive composites, Cold Reg. Sci. Technol. 86 (2013) 22–35.

[147] B. Han, K. Zhang, X. Yu, Enhance the thermal storage of cement-based composites with phase change materials and carbon nanotubes, J. Solar Energy Eng. 135 (2) (2013), 024505.

[148] B. Xu, Z. Li, Paraffin/diatomite/multi-wall carbon nanotubes composite phase change material tailor-made for thermal energy storage cement-based composites, Energy 72 (2014) 371–380.

[149] Y. Zhao, Y.X. Duan, W.W. Li, Z.F. Liang, The absorption properties of multiwalled carbon nanotube composites at 8mm wave band, Acta Mater. Compos. Sin. 3 (2007). 23–2.

[150] B. Wang, Z. Guo, Y. Han, T. Zhang, Electromagnetic wave absorbing properties of multi-walled carbon nanotube/cement composites, Constr. Build. Mater. 46 (2013) 98–103.

[151] D. Micheli, R. Pastore, A. Vricella, R.B. Morles, M. Marchetti, A. Delfini, V.M. Primiani, Electromagnetic characterization and shielding effectiveness of concrete composite reinforced with carbon nanotubes in the mobile phones frequency band, Mater. Sci. Eng. B 188 (2014) 119–129.

CHAPTER 6

Design, performance, and mechanism of cement-based materials with 2D nanomaterials

Wu-Jian Long
Guangdong Provincial Key Laboratory of Durability for Marine Civil Engineering, Shenzhen Durability Centre for Civil Engineering, College of Civil and Transportation Engineering, Shenzhen University, Shenzhen, Guangdong, PR China

6.1 Introduction

Numerous researchers have investigated the effect of nanoadditives (such as 0D nanoparticles and 1D carbon nanotubes, as discussed in previous chapters) on the microstructure and characteristics of cementitious composites during the last two decades. However, due to the low aspect ratio, 0D nanomaterials such as different nanoparticles are unable to restrain micro-cracks derived from nano-sized cracks, and hence are unable to improve the 'efficiency of reinforcement [1–5]. The 1D nanomaterials such as carbon nanotubes have strong Van der Waal interactions with the hydration products, but they tend to form agglomerates in the cement matrix [3,6–9]. It is graphene and its derivatives (mainly graphene oxide, GO, and reduced graphene oxide, rGO), which are 2D nanomaterials with a largest portfolio of outstanding characteristics, are gaining more interest for the next generation of cement technologies [10–17]. The 2D nanomaterials are not only similar to 1D nanomaterials in its attractive forces, but comprised sp^2 (hybridization involved the intermixing of one "s" with two "p" orbitals) bonded carbon atoms which bestow graphene with excellent mechanical properties [12]. Graphene-based nanomaterials exhibit potential to not only enhance mechanical strength and durability of cement-based construction materials but also provide smart functionalities duo to their remarkable electrical properties [18,19]. In this chapter, a comprehensive review on research progress of design, performance, and mechanism of cement-based materials with 2D nanomaterials, mainly graphene nanosheets and their derivatives, is conducted.

Nanotechnology for Civil Infrastructure
https://doi.org/10.1016/B978-0-12-817832-4.00001-8

6.2 Physical and chemical properties of 2D nanomaterials

Graphene is the fundamental structural unit for graphitic materials of any dimension. It is made out of a single-layer carbon atom sheet packed tightly into a two-dimensional honeycomb structure [11]. It can be rolled into 1D nanotubes, wrapped into 0D fullerenes, or stacked into 3D graphite [10], as shown in Fig. 6.1. GO, reduced graphene oxide (rGO), and graphene nanoplatelets (GNPs) are commonly used 2D graphene-based nanosheets to enhance cement-based materials. They can be fabricated using a variety of techniques, including epitaxial growth, ultrasonic exfoliations of graphite, chemical vapor deposition, and so on [20–23]. Over the last few years, significant progress has also been achieved in the manufacture of graphene and related materials by improving the quantity and quality, also via the tailoring of their properties [24,25].

Among all the different forms of graphene-based 2D nanomaterials, GO received the most attention in making novel cementitious composites. It consists of monolayer sheets with a hexagonal carbon network, but has hydroxyl, epoxide, carboxyl, and carbonyl functional groups on its surface and thus exhibits higher reactivity with cement and better dispersion in the matrix [26]. Compared to GO, rGO has less functional groups on its surface, which gives it intermediate properties between GO and graphene. GNPs

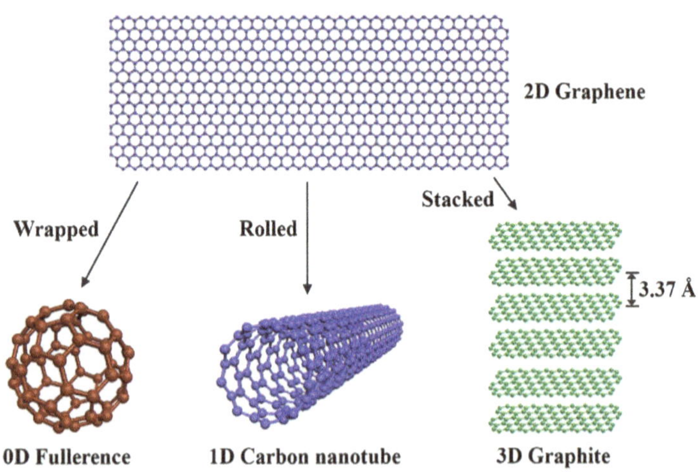

Fig. 6.1 Graphene is fundamental building block for graphitic materials, can be wrapped up into 0D fullerenes, rolled into 1D nanotubes or stacked into 3D graphite [10].

comprised layers of graphene sheets with thickness < 100 nm and diameter of several micrometers. Due to its increased thickness, GNPs are a cheaper alternative to graphene and therefore the preferred form for large-scale structural composites [27,28]. However, its smaller specific surface area compared to single graphene sheets decreases its reinforcement efficiency. The chemical representation of graphene and its derivatives are shown in Fig. 6.2.

In general, 2D graphene-based nanosheets have larger specific surface area, excellent elastic and tensile strength, high electrical, and thermal conductivity due to their unique physical and chemical properties. Most of the remarkable physicochemical properties of graphene and its derivatives arise from its 2D microstructure, which is comprised by adjacent carbon atoms linked by alternating single and double covalent bonds. This unique microstructure makes graphene the strongest, hardest, most impermeable, most stretchable material ever found [30]. In addition, graphene is chemically inert, presents excellent thermal conductivity and unique electrical and optical properties [31]. More importantly, they are easier to produce in large quantities with fully reproducible properties and easily dispersed in the cement matrix, which makes it attractive for various applications in construction materials [13].

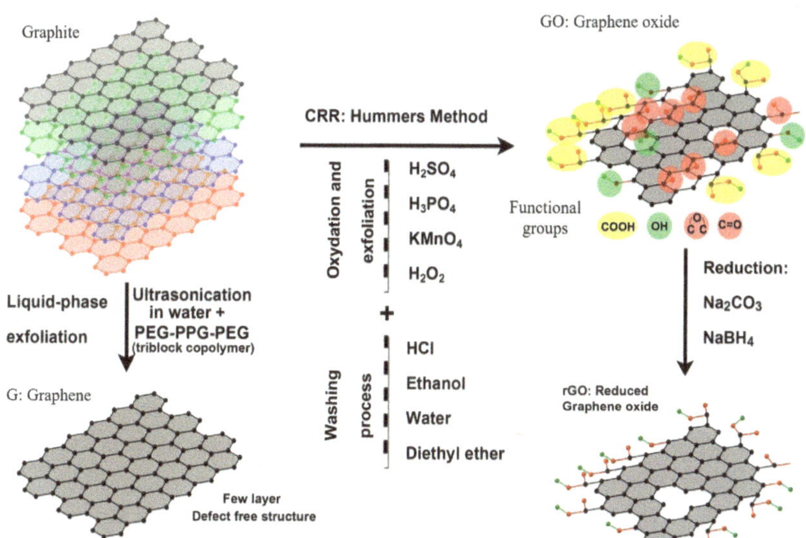

Fig. 6.2 Chemical representation of graphene and its derivatives [29].

6.3 Dispersion of 2D nanomaterials in cement-based materials

Existing research has shown that adding a modest amount of nanomaterials to cement-based materials can significantly improve their mechanical and microstructure properties [32–35]. An excessive amount of nanomaterials, on the other hand, does not always imply superior performance. The excessive use of nanomaterials in cement matrix induces non-uniform dispersion of the nanomaterials, high water consumption, and agglomeration due to their large specific surface area and high surface energy, especially 2D nanomaterials such as graphene-based nanosheet [4,36–38]. When 2D nanomaterials are non-uniformly dispersed in the cement matrix, their sizes may not be as close to those of the calcium silicate hydrate (C-S-H) gel, and thus the cement matrix at the nanoscale may not be enhanced [39]. Larger voids, weaker interfacial zones, and concentrated stresses in the cement matrix may all be the negative consequences of agglomerates [40–43].

Long et al. [44] investigated the effect of different two states of GO, namely uniformly dispersed, and re-agglomerated, on properties of cement paste. Two distinct dispersion states were achieved by altering the mixing sequence. The experimental results showed that the yield stress and plastic viscosity increased with the uniformly dispersed GO when compared to those of re-agglomerated GO cement paste. Moreover, the 3-day compressive and flexural strengths of uniformly dispersed GO paste were 8% and 27%, respectively, higher than those of re-agglomerated GO pastes. The results of X-ray diffraction, Fourier transform infrared spectroscopy, and scanning electron microscopy (SEM) analyses demonstrated that uniformly dispersed GO more effectively promotes the formation of hydration products in hardened cement paste. Furthermore, a porosity analysis using mercury intrusion porosimetry revealed that the homogeneous dispersion of GO can better inhibit the formation of large-size pores and optimize the pore size distribution at three and 7 days than the re-agglomerated GO. Thus, the uniform dispersion and distribution of 2D nanomaterials is crucial for the enhancements in mechanical and microstructural performance of cement composites.

At present, the uniform dispersion of 2D nanomaterials in cement-based materials is commonly achieved by combined physical and chemical methods. The physical dispersion methods include ultrasonication, ball milling, and high shear mill [43]. The chemical dispersion methods usually use surfactants for surface activation, such as polycarboxylate-based high-range water reducer (P-HRWR) and naphthalene-based high range water reducer (N-HRWR) [8,45,46]. Most studies focus on the dispersion of 2D nanomaterials in aqueous solutions [41,47,48]. However, a homogeneous dispersion of 2D nanomaterials in aqueous solutions may not be a guarantee of a uniform dispersion in cement pastes [43,49]. Due to the complex electrolyte, the dispersion and stability of 2D nanomaterials in the pore solution of cement matrix can be very different from those in aqueous solutions.

Vallurupalli et al. [50] found that the degree of dispersion of the GO in water improved with the use of and increase in sonication time. A maximum dispersion occurred by using an optimum dosage of HRWR corresponding to four times the mass of GO coupled with sufficient sonication for 48–56 min. Zhao et al. [51] studied the different dispersion behavior of GO in $Ca(OH)_2$, $CaCl_2$, and NaOH solutions. Their results showed that the prime factor responsible for the immediate aggregation of GO in cement paste is the complexation of Ca^{2+}, rather than the deoxygenation of GO in the alkaline environment.

Yan et al. [52] compared six different kinds of superplasticizers to optimize the dispersion of GO and found that polycarboxylate superplasticizer exhibits the best dispersibility with the optimal parameters found to be 30% ultrasonication energy with a SP-to-GO ratio of 1:1. In addition, polycarboxylate superplasticizer is capable of keeping dispersed GO separated even in strong alkaline environment. Lin et al. [53] modified GO using a tetraethyl orthosilicate (TEOS) to synthesize thin hybrid GO-SiO_2 nanohybrids (GOS). Their results demonstrated that GOS had significantly longer dispersion stability than GO. The authors proposed that the formation of a thin silica coating on GOS prevented cross-linking between Ca^{2+} ions from solution and carboxylate groups from GO. The thin silica coating also functioned as a spacer and minimized agglomeration caused by van der Waals forces. Compared to batches with GO, GOS nanohybrids provided additional 26% and 22% enhancement of compressive strength and 21% and 31% enhancement of tensile strength at 7 and 28 days, respectively.

Long et al. [54] also studied the effects of three different dispersing agents, including polycarboxylate-based high-range water reducer (P-HRWR), naphthalene-based high-range water reducer (N-HRWR), and air-entraining agent (AEA) on the dispersion of GO in aqueous solution, simulated concrete pore solution, and suspension of cement pastes. The results showed that the dispersion effect of GO in aqueous solutions was improved with different dispersing agents. As the cement content and pH of aqueous solutions increased, GO re-agglomerated and precipitated in an alkaline solution. The authors proposed a mechanism, as shown in Fig. 6.3, that electrostatic interactions and steric hindrance provided by

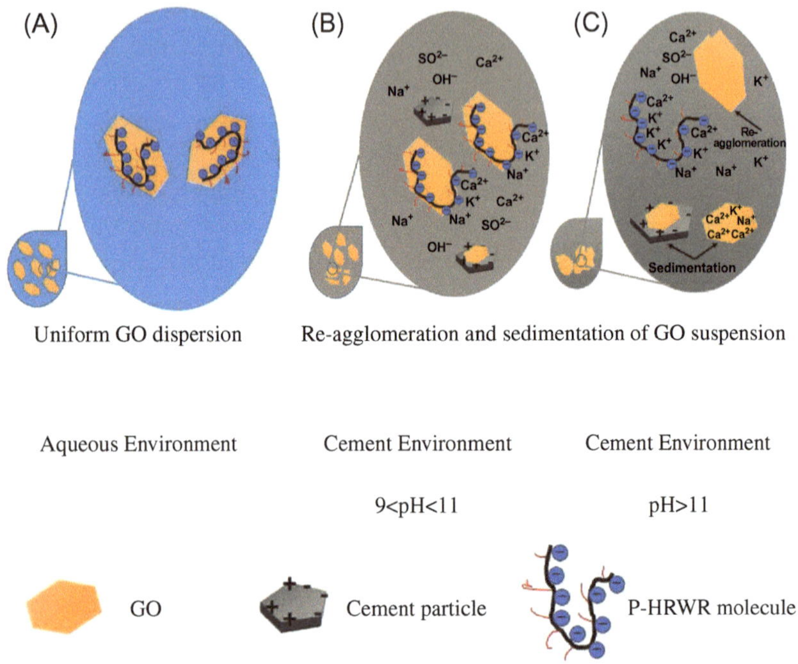

Fig. 6.3 Schematic representation of the mechanisms of destabilization of GO dispersion in (A) aqueous solution; (B) cement paste $9 < pH < 11$; and (C) cement paste $pH > 11$ for the P-HRWR, adopted from [54].

the P-HRWR further made GO stable in aqueous solutions. The ions and pH of cement pastes increased with the increasing amount of cement, which caused the separation of P-HRWR from GO. Therefore, GOs were re-agglomerated and absorbed on the surface of the cement particles, resulting in GO sedimentation.

In summary, the aggregation of GO occurs immediately when GO is introduced in cement paste due to the cross-linking effects of Ca^{2+}. The GO aggregates are not stable under high-shear mixing. They can be separated into small pieces mechanically. The use of polycarboxylate superplasticizer seems to be an effective method to disperse GO in cement pore solution. Close attention should be paid to the required dosage of polycarboxylate superplasticizer and the adding sequence of polycarboxylate superplasticizer, GO, and cement. It is imperative to develop new kinds of surfactants compatible with cement to disperse GO more effectively. Moreover, the deoxygenation process of GO with the presence of dispersing agents needs to be further investigated.

6.4 Performance of 2D nanomaterials in cement composites

6.4.1 Workability

The negative impact on the workability (rheological behavior), a critical property of fresh cement mixtures, is another key challenge of incorporating graphene-based nanosheets to cement composites [24]. For example, Qureshi et al. [55] assessed the workability of cement composites with GO or rGO using a mini-slump test. The slump diameters resulted in mini-slump test are presented in Fig. 6.4. Both the static and dynamic flow diameters were reduced approximately 29% and 16%, respectively, in the 0.06% GO composite, compared to the control mixtures (0% GO or rGO). The flow diameter decreased with the increasing content of GO, which indicates that the loss of workability is proportional to the percentage of GO in the composite paste. This is due to the hydrophilic high specific surface area and high interlayer distance (0.87 nm) of GO nanomaterials 2D plane, which requires extra water to wet their surface. The static and dynamic flow diameters in the rGO composites increased approximately 95% and 24%, respectively, compared to control mixture. A HRWA was used in the rGO solution for an efficient dispersion, which improved the workability similarly in all the rGO composites. Also, the rGO has a minimum impact on the workability of the composites. The rGO does not extract extra water from the composite since the rGO is almost hydrophobic with a low interlayer distance (0.36 nm).

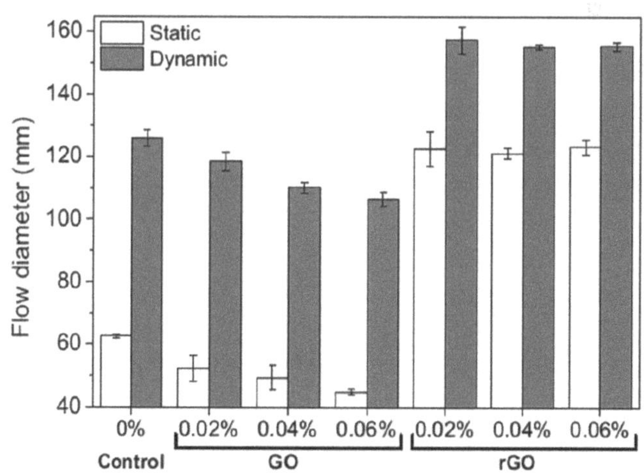

Fig. 6.4 Static and dynamic flow diameter in the mini-slump test [55].

Pan et al. [1] observed that the addition of 0.05% wc (weight of cement) of GO sheets can lead to a reduction of the workability by about 42% of cement pastes with w/c of 0.5, whereas Li et al. [56] and Shang et al. [37] observed, for the same GO content, a reduction of 25% for pastes with w/c of 0.4. Substantial reduction in fluidity was also observed for the GO-AAS (alkali-activated slag) mortars. Pan et al. [1] attributed the reduction in workability to the large surface area and hydrophilicity of GO nanosheets, which can absorb some of the free water in the fresh mixtures to wet their surface [57,58]. However, Shang et al. [37] later demonstrated that the negative effect of GO on the workability of cement paste is mostly due to the flocculation between the GO and cement grains by electrostatic interactions, which entraps large amount of free water. Similarly, Wang et al. [59] proposed that GOs are adsorbed onto the cement grains' surface and, due to the van der Waals attraction of the nanosheets, the cement particles can attach to each other, thus decreasing the fluidity of the mixture. In addition, the reaction between the –COOH groups in GO and metal ions in the hydration product occurs during absorption, producing $-COO^-$. This can result in chemical cross-linking of GO nanosheets and the formation of GO agglomerates, which subsequently reduces free water and the workability of cement paste due to their high water entrapment capacity [60,61].

To offset the negative effects of the GO on the workability of cement pastes, different approaches have been recently proposed. Wang et al. [62] proposed the use of fly ash in the GO-cement system. Based on the quantitative analysis of the paste's rheological parameters, the authors suggested a mechanism governing the positive influence of fly ash on the fluidity of the GO-cement paste. As shown in Fig. 6.5, the authors proposed that fly ash can interrupt the flocculation configurations of GO-cement paste, release the entrapped water and as result increase the mixture fluidity. The spherical shape, size gradation, and lower water requirement of the fly ash were considered as possible factors contributing to this phenomenon. Shang et al. [37] demonstrated that GO-encapsulated silica fume (GOSF) renders the cement paste a higher fluidity compared with cement pastes with only silica fume. This behavior was attributed to the synergetic effect of the surface activity of GO and the shape effect of silica fume. More recently, Wang et al. [63] observed that polyether amine functionalized-GO can increase the slump flow of cement pastes. The improvement in fluidity was attributed to the steric hindrance of the hydrophilic polyether amine chains.

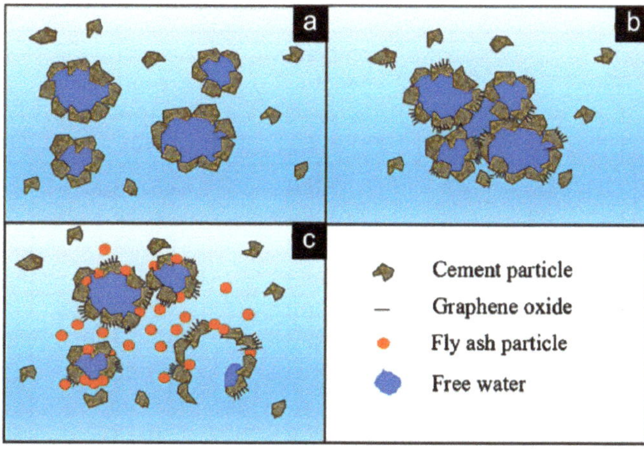

Fig. 6.5 Effect of GO and fly ash on formation of (A) cement paste, (B) GO-cement paste, (C) fly ash-GO-cement [62].

6.4.2 Mechanical properties

6.4.2.1 Static mechanical properties

The influence of 2D graphene-based nanomaterials on static mechanical properties of cement composites is presented in Table 6.1. It is clearly shown from the table that 2D nanomaterials can enhance both the compressive and flexural strength of cement composites significantly, with relatively small amount of addition. The outstanding reinforcement by adding GO is due to the possible promoted cement hydration, refined capillary pore structure, and reduced air void volume [71,82]. Although, it can also be seen from Table 6.1 that significant differences of the test results exist, even with almost the same mix design (same GO content and same w/b). The reinforcement efficiency is determined by many influencing factors including, the water/binder ratio, curing age, cement type, as well as dispersion, size, type, and content of nanomaterials. The mechanisms of GO reinforcement in cement-based materials have not been clearly figured out, to explain the conflicting data.

Duan et al. [1] proposed that the GO reinforcement is mainly attribute to the strong interfacial adhesion between GO and cement matrix, as shown in Fig. 6.6. They described a toughening mechanism for these composites similar to that reported earlier for graphene ceramic composites and GO-reinforced epoxy polymers. Unlike the plain matrix in which cracks usually traverse in a straight-through manner, in a graphene-cement

Table 6.1 Static mechanical performance of nano-reinforced cement-based materials.

Matrix	Type/ Content (wt%)	Water/ Binder	Compressive strength improvement (%)/age	Flexural strength improvement (%)/age	References
Paste	GO/0.02	0.3	60.1/28 d	84.5/28 d	[64]
Paste	GO/0.02	0.3	13.0/28 d	41.0/28 d	[65]
Paste	GO/0.03	0.29	42.5/28 d	55.0/28 d	[62]
Paste	GO/0.03	0.33	18.8/28 d	56.6/28 d	[66]
Paste	GO/0.04	0.3	28.6/28 d	43.2/28 d	[67]
Paste	GO/0.04	0.4	46.6/28 d	14.2/28 d	[68]
Paste	GO/0.05	0.29	52.4/3 d	90.5/28 d	[69]
Paste	GO/0.05	0.29	66.4/7 d	69.4/7 d	[70]
Paste	GO/0.05	0.5	59.0/28 d	33.0/28 d	[1]
Paste	GO/0.2	0.5	21/28 d	26/28 d	[71]
Mortar	GO/0.02	0.5	20/28 d	32/28 d	[72]
Mortar	GO/0.022	0.42	27/28 d	26/28 d	[46]
Mortar	GO/0.022	0.4	34.1/28 d	33/28 d	[73]
Mortar	GO/0.03	0.4	38.9/28 d	60.7/28 d	[74]
Mortar	GO/0.05	0.37	24.4/28 d	70.5/28 d	[69]
Mortar	GO/0.1	0.485	37.5/28 d	77.7/28 d	[75]
Mortar	GO/0.5	0.3	126.6/28 d		[76]
Mortar	GO/1	0.45	86/28 d		[77]
Concrete	GO/0.1	0.5	14.2/7 d	4.0/3 d	[78]
Paste	rGO/0.02	0.32	22.0/28 d	70.0/7 d	[79]
Paste	rGO/0.06	0.45	15/28 d	33.7/28 d	[80]
Paste	GNPs/0.05	0.35	24/28 d	8/28 d	[81]
UHPC	GNPs/0.3	0.2		59/28 d	[27]
UHPC	GO/0.01	0.2	7.8/28 d	11.9/28 d	[48]

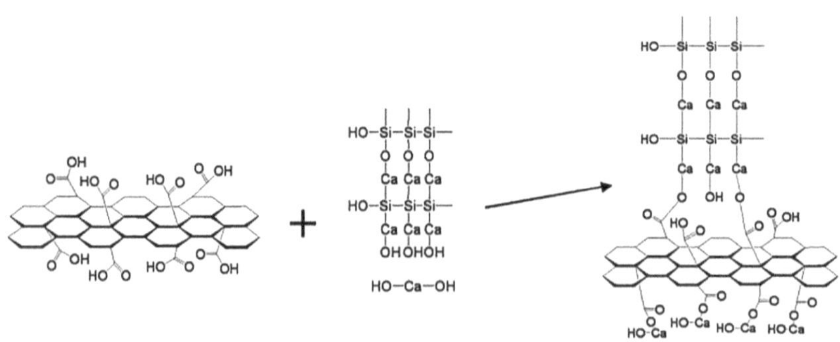

Fig. 6.6 Schematic of reaction between carboxylic acid groups and hydration productions ($Ca(OH)_2$ and C-S-H) of cement [1].

composite, cracks are arrested and cannot pass through the graphene sheets. GO sheets can block and divert the in-plane propagation of micro cracks.

Sharma et al. [77] studied the effect of size of GO nanosheets on mechanical properties of cement nanocomposites. It was found that the structural and mechanical properties of the composites are strongly dependent on the size of the incorporated GO sheets. For example, composites prepared using larger GO sheets (14 nm thick and 900 nm average size) improved the compressive strength by a maximum of 63%, while the same loading ratio of smaller and thinner GO nanosheets (3 nm thick and 100 nm average size) formed superior crystalline structures with a maximum strength improvement of 86%. Another study by Abrishami and Zahabi [83] demonstrated that incorporation of amine functional groups on GO exhibited positive effects on its reinforcing efficiency. Addition of NH_2-GO further reduced the porosity and improved the compressive and flexural strengths of mortar by 23% and 38%, respectively, compared to pure GO. Recently, graphene and its derivatives have also been utilized to further increase the mechanical strength of dense cement-based matrices, demonstrating the great potential of employing graphene nanomaterials to produce ultra-high-performance concrete (UHPC). For example, by increasing the content of GNPs (2–10 nm thick) by up to 0.3% wc, Meng and Khayat [27] found that the tensile strength and energy absorption capacity of UHPC was increased by 45% and 153%, respectively. Such improvement was attributed to a combination of "bridging effect" and "filler effect" of the GNPs. In addition, mixtures with high GNPs content presented a slip-hardening behavior at the post-cracking zone. Another study showed that UHPC mixtures with 0.01% wc of GO had an increase in compressive strength from 117 to 127 MPa and in flexural strength from 8.9 to 10.0 MPa.

Qureshi et al. [29] examines and compares properties of cement paste containing three forms of graphene-based 2D nanomaterials synthesized from epigenetic graphite deposit, namely, GO, reduced graphene oxide (rGO), and pristine GNPs (G). The 28-days compressive and flexural strengths of graphene (GO, rGO, and G) cement composite improved by 28% and 81%, 30% and 84%, and 39% and 38%, respectively, compared to the control mixture (cement paste without graphene materials).

6.4.2.2 Dynamic mechanical properties

Although cement-based materials are the most commonly used building materials in the world, in some cases, they may fail to satisfy engineering mechanical requirements due to the inherent brittleness, low tensile strength, and high probability of cracking. Moreover, most of the structural materials are unable to reduce vibrations to a sufficiently low level (Particle peak vibration ≤ 50 mm/s), and the dynamic mechanical properties of the traditional reinforced concrete structures are typically poor [84,85]. Cement-based materials modified with traditional chemical additives and mineral admixtures can still exhibit relatively low storage modulus, chemical stability, and durability. Hence, it is desirable to develop a novel cement material characterized by superior energy dissipation capacities, high storage modulus, and good seismic properties. 2D nanomaterials are promising in enhancing the dynamic mechanical properties of cement-based materials. Due to their good shear deformation properties (similar to those of rubber and CNTs), the addition of graphene-based nanosheets in cement matrix can result in a high value of loss factor (the ratio of the average power loss to the peak load loss, during a specified period of time), while the stiff cement matrix ensures a high storage modulus [86]. Long et al. [71] investigated the effect of GO nanosheets on the dynamic mechanical properties of cement paste. The results of dynamic mechanical testing revealed that loss factors of the pastes containing 0.05, 0.10, and 0.20 wt% of GO were improved by 31%, 58%, and 77%, respectively. The maximum storage modulus of 52% was observed at a GO content of 0.1 wt%. The authors concluded that the increased dynamic mechanical properties of cement paste with the addition of GO are due to the large internal surface area, moderate porosity, and non-uniform stress distribution. The proposed mechanism is illustrated in Fig. 6.7.

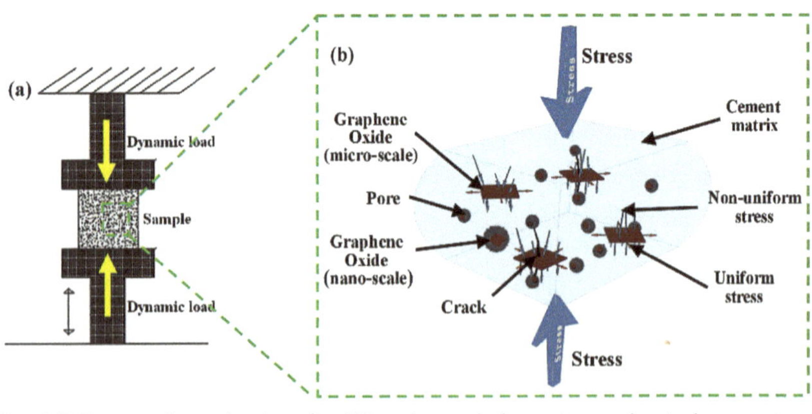

Fig. 6.7 Proposed mechanism for GO enhanced dynamic mechanical property of cement paste [71].

6.4.3 Hydration and microstructural characteristic

Several studies concluded that 2D nanomaterials, especially GO, have significant effect on hydration and bonding of cement matrix, which leads to the drastic improvement in macro and micro properties of the hardened cement composite. However, there are still controversial about the explanation. Lv et al. [24] proposed the template effect that GO was regarded as a substrate and catalyst for flower-like crystals formation. It was suggested that the process of hydration and formation of crystals were controlled by the oxygen-containing groups (–$COOH^-$, –OH^-, –SO_3^-) on the surface of GO as these active groups would easily react with the active groups of cement hydration products. The authors proposed that the GO would preferentially absorb C_3S, C_2S, C_3A, and C_4AF due to its active functional groups, resulting in the nucleation and growth of crystals on the GO surface. The conclusion was then questioned by Cui et al. [87], who proposed that the flower-like crystals may be calcium carbonate crystals from carbonation during sample preparation, not cementitious hydration products.

Lin et al. [88] proposed that GO can play the role of crystal nucleus for cement hydration, which leads to the acceleration of the hydration of cement. However, the authors stated that GO could absorb water promoting the chemical reaction at the mid-stage of the hydration process. Authors suggested that the acceleration of hydration was mainly due to the oxygen containing function groups on GO, which would provide platforms for water and cement to continue the crystal nucleation process, so the hydration process can be further enhanced. However, Wang et al. [59] argued that GO nanosheets would be absorbed on the cement surface similar to the case of water reducing agent. Due to the Van der Waal's force, the layers of GO would evoke a negative influence on the fluidity of cement. Wang et al. [66] also showed that the carboxyl functional groups of GO can react with Portlandite (CH) to generate 3D network structure. Ghazizadeh [89] investigated the effect of GO on the hydration of alite, and found that GO only accelerates the hydration of alite marginally, and can easily aggregate in alite paste. In fact, due to heterogeneity of cement-based materials and various oxygen functional groups, the chemical mechanism in GO cement-based composites still remains unclear.

Long et al. [82] conducted a comprehensive characterization of the microstructure and micromechanical properties of GO cement-based

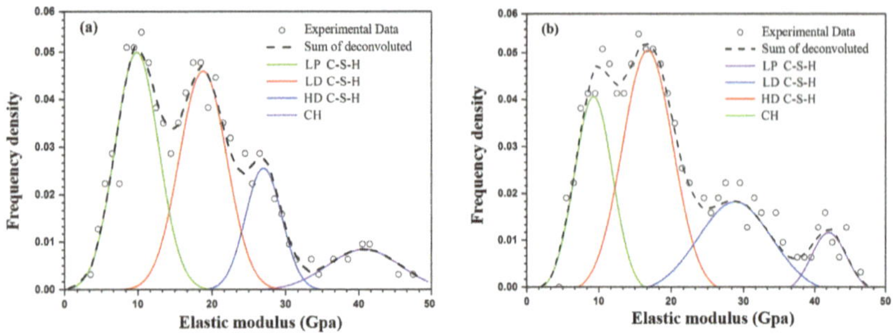

Fig. 6.8 The modulus PDF of cement paste with different GO content at 0.5 w/c (A) OPC at 28 days and (B) with 0.1% GO addition at 28 days [82].

materials. The micro-mechanical properties of each phase hydrated products with GO addition were investigated by the synergistic analysis of nanoindentation, SEM equipped with energy-dispersive spectrometer (EDS), and backscattered electrons (BSEs). The authors found that the addition of GO obviously enhanced the elastic modulus and the compressive and flexural strengths of hardened cement paste at 28 days. Four phases can be identified and quantified in the hardened cement paste made with 0.5 w/c ratio using nanoindentation. The volume fraction of loosed-packing calcium silicate hydrate phase (LP C-S-H) was significantly reduced with the addition of GO, while the volume fraction of high-density calcium silicate hydrate phase (HD C-S-H) was increased compared to ordinary Portland cement (OPC), this indicate that GO had proportioning effect of hydrated product, as shown in Fig. 6.8.

The authors also found that micrometer and sub-micrometer scaled cracks were inhibited in hardened cement paste with the addition of GO. The SEM-BSE results, as shown in Fig. 6.9, confirmed that the CH phase was increased with the use of 0.1% GO addition, indicating that GO can promote the cement hydration. The pore phase was also notably reduced with the use of 0.1% GO addition. The combined use of nanoindentation and SEM-BSE observation further identified the microstructure and micromechanical properties of GO cement-based materials. The synergistic analysis of microstructure and micromechanical properties at multiscale provides valuable information toward the material design of cement-based materials with 2D nanomaterials.

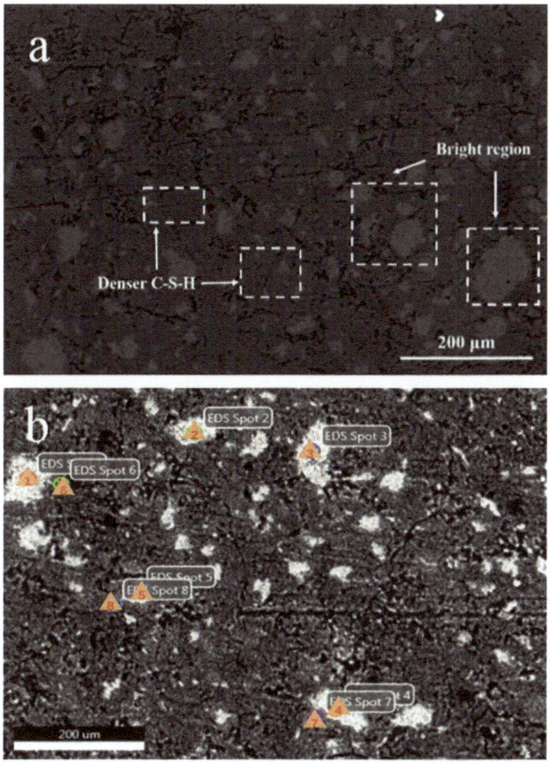

Fig. 6.9 Microstructure under (A) SEM-BSE and (B) EDS spots of hardened cement containing 0.1% GO addition, the fractions of carbon atom of the bright region were higher [82].

6.4.4 Durability

Cement-based materials are susceptible to various forms of deterioration, including carbonation, chloride induced corrosion, sulfate attack, alkali silica reaction (ASR), calcium leaching, freezing, and thawing cycles, which can severely shorten their service life. Concrete structures exposed to the aggressive agents are mainly governed by their pore structure and permeability. According to a recent study, introducing a small amount of nanomaterials can significantly improve the durability of concrete by optimizing the pore structure and changing the transport properties of the cement matrix. However, the number of studies is still limited, and more in-depth research into the reinforcement mechanisms of the influence of graphene-based nanosheet on long-term properties of concrete is required.

6.4.4.1 Carbonation

Carbonation of cement-based materials is a serious problem and considered a neutralization phenomenon inside hardened cement-based materials. The process of carbonation involves the reaction between the carbon dioxide and calcium-containing phases, which reduces the pH of the pore solution, thus leading to de-passivation of reinforcing bars and corrosion in the presence of moisture and oxygen [90]. Among various nanomaterials, GO and GNPs represent unique 2D nanomaterials with a sheet-like structure that can effectively improve the strength and durability of cement-based materials and ensure a relatively large contact area with the host matrix. Furthermore, the carboxylic functional groups of GO can form strong covalent bonds between the GO and C-S-H or CH. Long et al. [91] studied the carbonation mechanism of cement paste containing GO. The obtained electrochemical impedance spectroscopy (EIS) results indicated that the ion diffusion and transport resistance increased after the incorporation of GO, while TGA results revealed that the carbonation of portlandite (CH) and calcium-silicate-hydrate (C-S-H) was significantly inhibited during early ages of carbonation due to the increased degree of hydration, as shown in Fig. 6.10.

In addition, the hybrid GO/hydration products from the carbonation process were characterized, and the formation of a hydrated phase coated with a carbonated layer was observed via SEM equipped with EDS. The porosity of the OPC decreased more significantly during the initial carbonation period as compared to the effect observed for the GO cement-based material. A hypothesis is proposed to explain the carbonation mechanism of cement paste with GO. Initially, CO_2 dissolution occurs inside the pores, which are around the face exposed to gaseous CO_2. In this stage, the diffusion and transport of CO_3^- ions are significantly hindered, mainly due to the increased extent of hydration and the denser structure with GO addition, which can effectively mitigate the dissolution of CH and decalcification of C-S-H. During this stage, a hydrated phase coated with a carbonated layer and GO/C-S-H or GO/CH composites are formed, which also increases the carbonation resistance. At longer carbonation times, both the dissolution of CH and decalcification of C-S-H are accelerated despite the more severe carbonation of the cement-paste specimen without GO. This is because the constant consumption of CH reduces the pH of the made cement system, which accelerates the carbonation of the cement paste. Since the CO_2 concentration on the exposed surface is relatively high, the carbonation

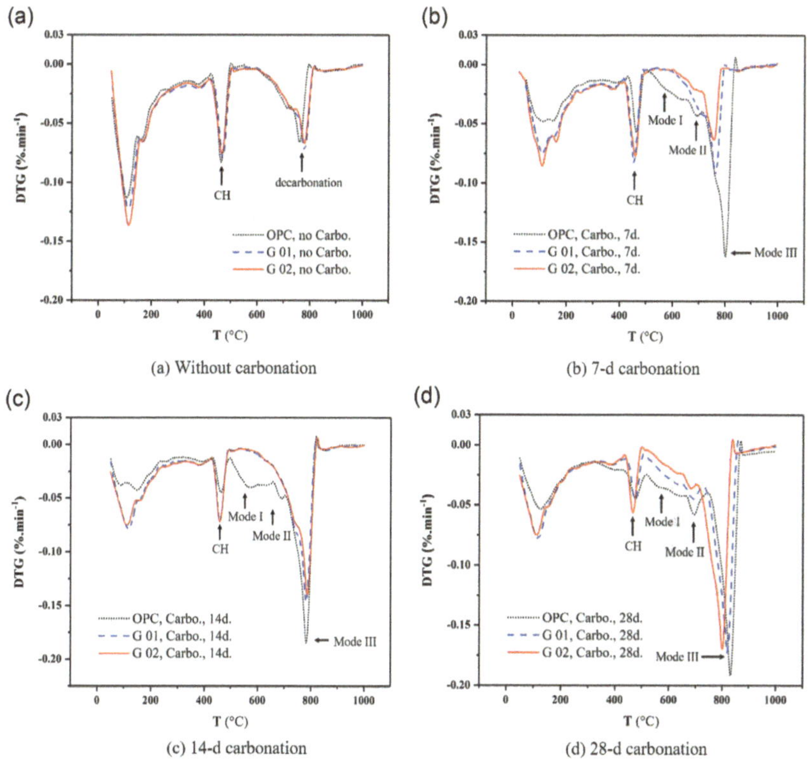

Fig. 6.10 The dissolution of CH, and especially decalcification of C-S-H, are suppressed during early carbonation stages with GO addition [91].

rates of the cement pastes made with and without GO can decrease with decreasing degree of CO_2 permeation.

Mohammed et al. [92] also found that mortars made with GO, after exposure of 18 months in a carbonation chamber, displayed a fivefold decrease in carbonation depth compared to the reference sample. Binding of Ca^{2+} ions on the surface of GO was also suggested as one of the reasons for durability enhancement.

6.4.4.2 Chloride ingress

Mohammed et al. [93] reported that mortars with 0.01% wc of GO presented a fivefold reduction in chloride penetration compared to plain mortars, as shown in Fig. 6.11. Furthermore, addition of 0.03% wc of GO in mortars decreased their initial water sorptivity ratio from 0.00046

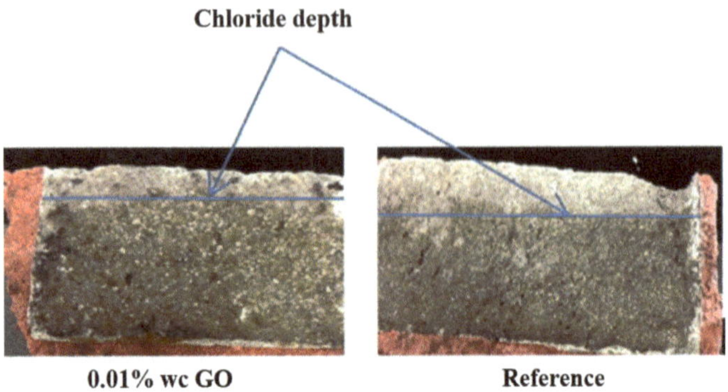

Fig. 6.11 Chloride front of the reference mortar compared to mortar containing 0.01% wc GO. Chloride penetration was approximately 26mm for reference sample, whereas only 5mm for GO-composite [93].

to 0.00020 $mm\,s^{-1/2}$. The authors suggested that the layered structures of GO may trap and reduce the chloride ingression depth.

Du et al. [40,94] carried out experimental analysis on the transport properties (chloride and water) of concrete containing up to 7.5% wc of graphene-based nanosheets at 0.5% wc increment. Concrete containing 1.5% wc graphene-based nanosheets exhibited the optimum performance with a reduction in water penetration depth, chloride diffusion, and migration coefficients by 80%, 80%, and 37%, respectively. The enhancement in the transport properties was mostly attributed to the increase in tortuosity and also to the pore refinement. Due to limitations in dispersing the graphene-based nanosheets, further improvement could not be achieved by increasing the graphene-based nanosheets content. Graphene-based nanosheets had also been applied to fabricate graphene-cement anodes for electrochemical chloride extraction (ECE) [95,96]. Zhao [95] synthesized an anode with graphite powder and cement with weight ratio 1:1. The anode exhibited an average ECE efficiency of 71% at electric current of 4 A/cm^2, which was slightly higher than that of the conventional Ti-RuO_2 mesh anode.

Through chloride diffusion tests, Zhao et al. [97] found that a small dosage of GO (0.011 wt%) can significantly reduce the content of chloride diffusion, and the average chloride depth was dropped by 29% compared to the plain sample. The authors also performed molecular dynamics simulations, and the results showed that the "caging effect" and "immobilizing effect" of GO sheet can inhibit the chloride ingress.

6.4.4.3 Calcium leaching

In addition to the chloride ingress, calcium leaching of cement-based composites is a common durability problem. It generally occurs in construction elements that on in contacted with aqueous environments with a pH lower than 12.5 over long periods of time such as water pipes, cisterns, dams, and nuclear waste storage facilities [98,99]. Up to date, graphene-based nanomaterials have proved to enhance the leaching resistance of cement-based composites. Muthu et al. [5] applied 0.5 mol/L HNO_3 solution as on accelerated of leaching method and reported that rGO addition can decrease the amount of bound water and capillary pores (10 nm–10 μm) in cement-based composites by up to 6% and 46% than the controlled pastes after exposure to calcium leaching. Long et al. [100] applied deionized water and indicated that GO addition also can contribute to the improvement of leaching resistance in cement-based composites. After leaching for 28 days, mass loss of CH in GO-cement composites decreased from 4.9% to 3.9% compared to the plain pastes, while the increment of total porosity mainly deriving from the increment of pores (1–5 μm) reduced from 25.6% to 3.8%. Furthermore, Long et al. [101] applied 6 mol/L NH_4Cl solution and explained that this improvement can be attributed to the absorption of GO nanosheets to Ca^{2+} ions in cement pore solution and the improvement of the microstructure in GO-cement composites. In addition, EIS was applied to characterize the calcium leaching process of GO-cement composites, and accurately predict the leaching depth of the composites.

6.4.4.4 Sulfate attack

Jiang et al. [102] investigated the influence of GO on the sulfate corrosion resistance of cement-based materials. The mechanical strength of specimens showed a continuous increase with the soaking time increasing to 90 d. Afterwards, the specimens showed a reduction in mechanical strength with the soaking time increasing to 135 d. The strength of specimens soaked for 135 d still showed higher performance compared with that of the specimens before the initiation of sulfate solution soaking. The authors proposed that GO could improve the resistance to SO_4^{2-} permeability of cement paste by preventing the formation of intumescent minerals in the mortar.

6.4.4.5 Freeze-thaw resistance

The freeze-thaw resistance of concrete can also be increased by incorporation of graphene-based nanosheets. Mohammed et al. [103] observed that mortars containing 0.06% wc of GO presented a weight loss of about

0.25% after 540 freeze-thaw cycles compared to about 0.8% for the reference sample. Mortars mixtures made with GO also displayed less scaling damage on the surfaces of GO. In addition to the higher compressive strength, the enhancement of freeze-thaw resistance of mixtures made with GO was associated with the increase of nanopores over mesopores, as less temperature is required to freeze the water in smaller pores [104]. The generated nanopores also created rooms for osmotic pressure release, hindering the propagation of microcrack and limiting the frost damage. Tong et al. [105] revealed that different types of graphene-based nanosheets exhibit completely different effects on frost resistance of mortar. For example, compared to the benchmark samples, pristine graphene-based nanosheets exacerbated the mortar's freeze-and-thaw resistance, whereas their oxidized form (GNOPs) enhanced it. The discrepancy was attributed to different pore pressure amplitudes as the water migrates from gel pores to capillary pores in the C-S-H during cyclic freeze/thaw loads. According to Grand Canonical Monte Carlo (GCMC) and MD simulation, the pore pressure amplitude in graphene-based nanosheets-reinforced C-S-H gels were 15.8 MPa, compared to 12.5 MPa for GONPs-reinforced C-S-H gel.

In summary, the incorporation of 2D nanomaterials in cement composites can block the transport of aggressive agents within the cement paste and improve the resistance to carbonation, frost and calcium leaching of cement composites. Compared to mechanical properties, there is far less information available on durability-related properties of cement composites containing 2D nanomaterials. Therefore, more in-depth studies are required to focus on the durability of 2D nanomaterials reinforced cement composites.

6.4.5 Functionality

Nanomaterials have high production costs compared to conventional concrete mix, and they require sophisticated protocols to assure correct dispersion within the cement matrix, limiting their use in real scale concrete structures to a large extent. The real promising added value of nanotechnology in concrete industry is represented by the unique and smart functions brought by 2D nanomaterials, such as self-sensing, self-cleaning, stabilization/solidification (S/S), anticorrosion, and so on [18,106].

6.4.5.1 Self-sensing

Due to its electrical conductivity and piezo resistivity-based strain sensing ability, graphene-based nanosheets can promote self-sensing capability of cement composites [106–109]. Saafi and co-workers [109] investigated the electrical and piezoresistive behavior of the geopolymers containing in situ reduced GO nanosheets. By increasing the content of rGO from 0 to 0.35% wc, the electrical conductivity of the composite improved from 0.77 to 2.38 $S m^{-1}$. The use of rGO also increased the gauge factor by up to 112% and 103% for specimens subjected to tension and compression, respectively. The rGO's electrical conductivity has also been exploited to create cement paste sensors, which can measure the chloride ion penetration in mortars independently of RH [110]. Pang et al. [106,107] noted that the increasing addition of graphene-based nanosheets considerably reduces the electrical resistance of OPC mortars. According to the authors, as the addition exceeds 3.6% of graphene-based nanosheets (by volume of mortar), the conductive nanomaterial can create a continuous electrical path along the specimen, and the composite's electrical conductivity became insensitive to its moisture content, enabling reliable damage measurements in actual applications.

Sun et al. [111] reported that, although cementitious composites with graphene-based nanosheets contents lower than 2% (by volume of composite) showed no obvious piezoresistive effect, composites containing graphene-based nanosheets between 4% and 10% exhibited 6.7%–15.6% of fractional change in electrical resistivity when the compressive stress was 20 MPa. Similar behavior was reported for multi-layer graphene [112]. Recently, Rhee et al. [113] observed that the electrical response of cement mortar filled with rice-husk graphene (GRH) presented moderate range of the electrical performance for GRH composite, a range similar to that of carbon fiber. However, although graphene-based nanosheets can increase the electrical conductivity of cement composites, graphene-based nanosheets-composites have been reported to exhibit inferior self-sensing capabilities (i.e., piezoelectric properties, gauge factor, strain sensitivity, and signal quality) compared to composites composed by other carbon materials [9,114]. Although the reasons have not been explored in-depth, one possible explanation for the inferior self-sensing performance of graphene-based nanosheets compared to other carbon nanomaterials could be associated with their dimensionality, as randomly aggregated thin sheets may not form continuous conductive pathways as effectively as 1D cylindrical- and 0D particle-shaped materials.

6.4.5.2 Heavy metal stabilization/solidification (S/S)

In recent years, a dramatic increase of electronic waste (e-waste) has led to global environmental concerns because of its complex and hazardous nature

[115]. The treatment of heavy metal waste by stabilization/solidification (S/S) with cement-based materials has been shown to be a sustainable option. Although waste materials are generally reused to replace the raw materials of the modern concrete, the use of waste materials may lead to secondary environmental pollution and may affect the eco-performance of the host concrete material. Recently, several researchers have attempted to apply nanotechnology to cement-based materials to improve their resistance to ion transport [5,116,117]. Long et al. [118,119] studied the effect of GO in inhibiting lead leaching of waste cathode ray tube (CRT) glass within cement-based materials. Lead leaching was evaluated using the toxicity characteristic leaching procedure (TCLP) and EIS. The TCLP results indicated that Pb leaching of waste CRT mortar with GO addition was significantly controlled below the regulatory limit (5 mg/L) specified by the United States Environmental Protection Agency (USEPA) and the Chinese Standard. The electrochemical parameters revealed that the microstructure of the bulk cover and waste CRT mortar interface was improved by adding GO, indicating that lead transfer within the bulk cover and waste CRT mortar interface was inhibited, as shown in Fig. 6.12. Better attachment between waste CRT glass/bulk paste and higher density hardened structure with 0.1 wt%

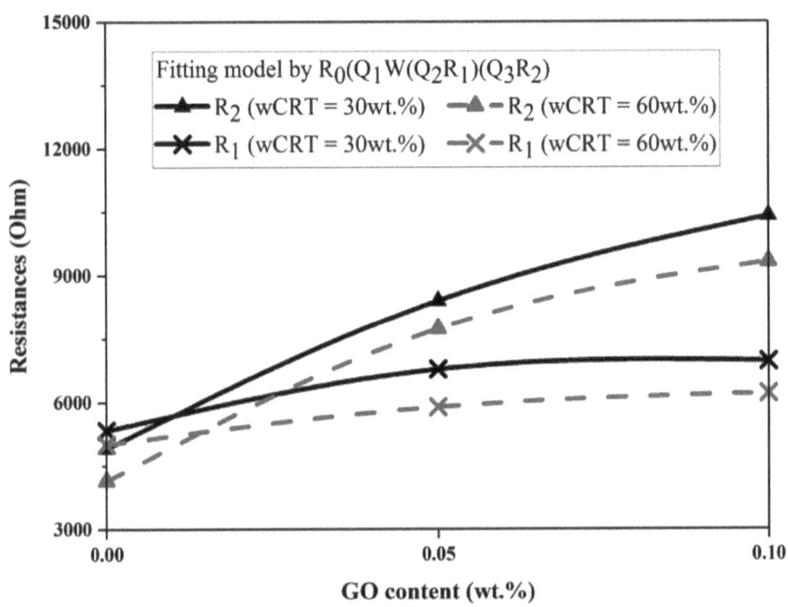

Fig. 6.12 Evolution of electrochemical parameters R1 and R2 vs GO content (0.00%, 0.05% and 0.10 wt%) with different waste CRT glass contents of 30 and 60 wt% [118].

GO were observed in SEM images, thus indicating that the hardened structure can effectively encapsulate lead ions out of the waste CRT glass in the interface and bulk paste. Their results show that GO has significant effect on inhibiting lead ions transfer. This solution has potential to improve the binding of contaminants originating from waste materials.

6.4.5.3 Electromagnetic shielding

With technological development, electronic devices extensively used in commercial and military applications are growing fast in number, continue to generate severe electromagnetic interference (EMI) and radiation. The substantial increase in EMI can result in malfunctioning and degradation of electronics [120,121], while the severe electromagnetic pollution can lead to harmful effects on the health of human exposed to electromagnetic fields [122]. GO, has been reported as a unique carbon nanomaterial with sheets-like structure for effectively improving the macro-properties of cement-based composites, as mentioned before. Due to its dominant shielding mechanism of electromagnetic radiation absorption, GO sheets have been reported to be effective for improving the shielding effectiveness of cement-based composites [123]. Besides, the sheet-like structure of graphene can create a lot more contact area with the host matrix, and furthermore the oxygen functional groups of GO make it exhibit a good interfacial attachment [124,125]. By introducing GO sheets onto the surface of carbon fiber, GO-deposited carbon fiber exhibits better interfacial attachment, which has been found more effective than carbon fiber in providing EMI shielding of cement-based composites. Long et al. [19] studied the coupling effect of GO addition (up to 0.10 wt% of cement) and waste CRT glass replacement for fine aggregates (30 and 60 wt%) in cement-based composites on the mitigating EMI. The electric permittivity was used in their study to evaluate the EMI shielding capacity of cement-based composites. The DC electrical resistivity of specimens is shown to increase insignificantly with increasing in GO content, but it increased remarkably with increasing the CRT glass content from 30 to 60 wt%. The 60 wt% replacement of waste CRT glass containing 0.1 wt% GO addition is shown to increase the relative permittivity by about 50% and 200% when the frequency is in the ranges of $10^{4-5} \times 10^6$ Hz and $10–10^3$ Hz, respectively, as shown in Fig. 6.13. They concluded that a significant increase in the permittivity can be obtained owing to the synergetic interaction between waste CRT glass and GO. The material with high value of the permittivity can strongly interact with electromagnetic radiation, which is preferred by EMI shielding.

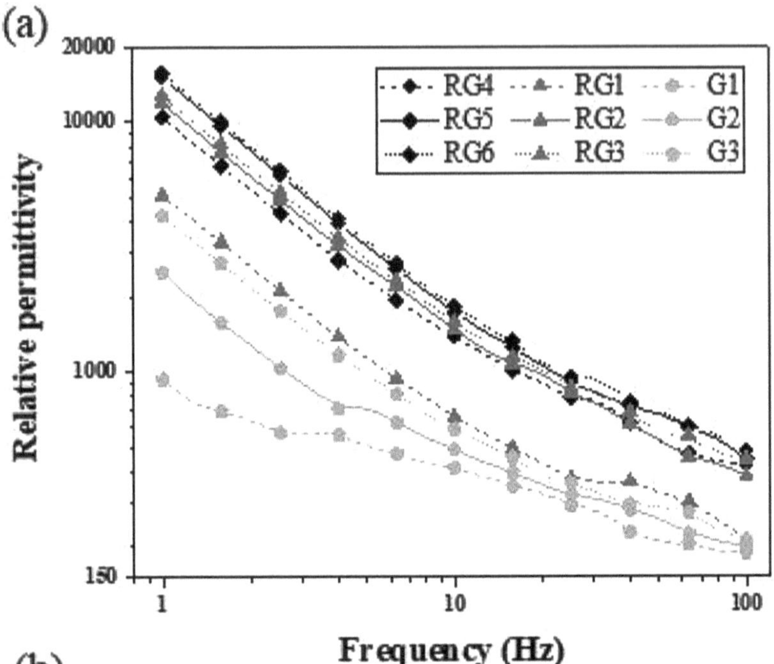

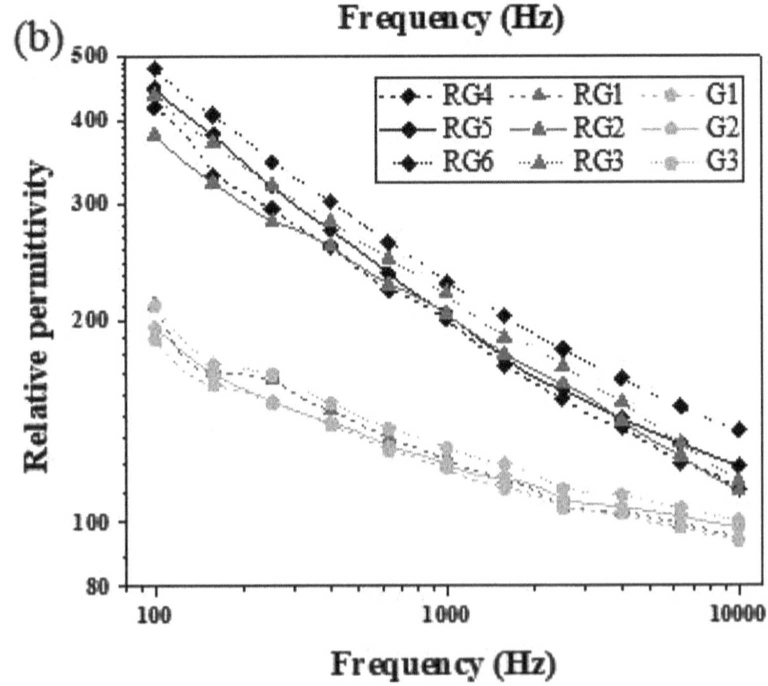

Fig. 6.13 Relative permittivity of waste CRT recycling (0%, 30%, and 60%) mortar at different GO contents (0%–0.1%) in the frequency ranging from: (A) 1–100 Hz; (B) 100–10,000 Hz [19].

The improvement in the EMI shielding ability of cement-based composites not only enables the application of these composites in mitigating electromagnetic pollution but also promotes large-volume recycling of toxic waste CRT glass.

6.4.5.4 Thermal properties

Thermal properties of graphene-based nanosheets have been exploited to directly increase the thermal conductivity of cement composites, which can improve their inner heat dissipation and lessen the effect of thermal cracking. By using time-domain thermoreflectance method, Sedaghat et al. [126] reported that incorporation of 5% and 10% wc of graphene can increase the thermal diffusivity (cm^2/s) of cement pastes by 25% and 75% at the first 44 h of curing. By quasi-steady-state method, Cui et al. [127] observed that 5% vc loading of nano graphite platelets (thickness = 1–5 nm) increased the thermal conductivity of mortar composites from 1.1 to 1.95 W/(m·K), while decreasing their specific heat by 17%. More recently, Chu et al. [128] observed that 0.01% wc of graphene sulfonate nanosheets can increase the thermal diffusivity of sacrificial concrete exposed to temperatures from 25 to 1000°C and improve the residual mechanical strength of the composites. These results indicate that graphene can function as a heat sink in cement composites, although its effectiveness in preventing thermal cracking still needs to be confirmed. Jing et al. [129] studied the effects of rGO on the thermal properties of cement composites. Results showed that the thermal conductivity and thermal diffusivity coefficient can be enhanced by 7.8% and 29%, respectively, by doping 1.2 wt% rGO. The enhanced thermal performance efficiently improves the heat propagation, and consequently reduces the temperature differences between the center and edge of the mortar at different layers.

6.4.5.5 Fiber reinforcement

Lu et al. [130] investigated the effect of GO on strengthening the interface bond of polyethylene fiber (PE)/cement matrix. In their study, GO was coated on PE fiber to strengthen its bond to matrix. Compared with PE-Mortar, GO/PE-Mortar showed a significant enhancement in first cracking strength, ultimate tensile strength, and strain by approximately 12%, 46%, and 70%, respectively. The authors explained that the hydrogen at honeycomb lattice of GO and oxygen in the hydroxyl and epoxy of GO can interact with oxygen in calcium silicate hydrogen (C-S-H) and hydrogen in PE chains via intermolecular H-bonding, contributing to strengthen the bond of PE fiber/matrix.

6.5 Summary and conclusions

Recent research studies demonstrate that graphene-based 2D nanomaterials with extraordinary properties have a great potential to fabricate high-performance cement-based composites. Compared to 0D and 1D nanomaterials, graphene and its derivatives have large aspect ratio along with their atomic thickness, providing large surface area for the nanosheets or their functional groups to interact with the cement matrix through the bridging and nucleation effects, thus contributing to accelerated cement hydration, uniform microstructure, and improved mechanical properties of graphene-cement composites. However, the reinforcing mechanism and exact relationship between graphene-based 2D nanomaterials and hydration products is not fully understood. Further in-depth and rigorous investigation is required. Also, the development of an efficient method for obtaining the appropriate and homogenous dispersion of graphene and derivatives thereof within cement matrix is highly desired as well as precise control over size and thickness of graphene sheets remains challenging.

Owing to its low-cost production, easy processability, and high dispersibility in water, GO has been the most studied 2D nanomaterial using in cementitious composites, whereas limited works have considered adopting other graphene related materials such as rGO and GNPs. The beneficial effect of GNPs on microstructure and mechanical properties of cement materials is highly restricted due to their large size. As an alternative to GO, the beneficial effect of rGO on the performance of cement composites is also strongly inhibited by the reduced number of oxygen functional groups. However, GNPs, rGO, and other graphene derivatives should gain more attention in developing concrete nanotechnology due to their remarkable electrical conductivity offering possibility for smart and functional cement-based structural materials.

References

[1] Z. Pan, L. He, L. Qiu, A.H. Korayem, G. Li, J.W. Zhu, F. Collins, D. Li, W.H. Duan, M.C. Wang, Mechanical properties and microstructure of a graphene oxide–cement composite, Cem. Concr. Compos. 58 (2015) 140–147.

[2] S.C. Paul, A.S. van Rooyen, G.P.A.G. van Zijl, L.F. Petrik, Properties of cement-based composites using nanoparticles: a comprehensive review, Construct. Build Mater. 189 (2018) 1019–1034.

[3] K. Chintalapudi, R.M.R. Pannem, An intense review on the performance of graphene oxide and reduced graphene oxide in an admixed cement system, Construct. Build Mater. 259 (2020), 120598.

[4] E. Ghafari, S.A. Ghahari, Y. Feng, F. Severgnini, N. Lu, Effect of zinc oxide and Al-zinc oxide nanoparticles on the rheological properties of cement paste, Compos, B 105 (2016) 160–166.
[5] M. Muthu, M. Santhanam, Effect of reduced graphene oxide, alumina and silica nanoparticles on the deterioration characteristics of Portland cement paste exposed to acidic environment, Cem. Concr. Compos. 91 (2018) 118–137.
[6] A. Cwirzen, K. Habermehl-Cwirzen, V. Penttala, Surface decoration of carbon nanotubes and mechanical properties of cement/carbon nanotube composites, Adv. Cem. Res. 20 (2) (2008) 65–73.
[7] S. Musso, J.M. Tulliani, G. Ferro, A. Tagliaferro, Influence of carbon nanotubes structure on the mechanical behavior of cement composites, Compos. Sci. Technol. 69 (11) (2009) 1985–1990.
[8] S. Parveen, S. Rana, R. Fangueiro, M.C. Paiva, Microstructure and mechanical properties of carbon nanotube reinforced cementitious composites developed using a novel dispersion technique, Cem. Concr. Res. 73 (2015) 215–227.
[9] D.Y. Yoo, I. You, S.J. Lee, Electrical properties of cement-based composites with carbon nanotubes, graphene, and graphite nanofibers, Sensors 17 (5) (2017) 1064–1076.
[10] S. Zhao, Z. Zhao, Z. Yang, L. Ke, S. Kitipornchai, J. Yang, Functionally graded graphene reinforced composite structures: a review, Eng. Struct. 210 (2020), 110339.
[11] K.S. Novoselov, A.K. Geim, S.V. Morozov, D. Jiang, Y. Zhang, S.V. Dubonos, I.V. Grigorieva, A.A. Firsov, Electric field effect in atomically thin carbon films, Science 306 (5696) (2004) 666–669.
[12] H. Yang, H. Cui, W. Tang, Z. Li, N. Han, F. Xing, A critical review on research progress of graphene/cement based composites, Compos. A 102 (2017) 273–296.
[13] E. Shamsaei, F.B. de Souza, X. Yao, E. Benhelal, A. Akbari, W. Duan, Graphene-based nanosheets for stronger and more durable concrete: a review, Construct. Build Mater. 183 (2018) 642–660.
[14] Y. Xu, J. Zeng, W. Chen, R. Jin, B. Li, Z. Pan, A holistic review of cement composites reinforced with graphene oxide, Construct. Build Mater. 171 (2018) 291–302.
[15] A. Anwar, B.S. Mohammed, M.A. Wahab, M.S. Liew, Enhanced properties of cementitious composite tailored with graphene oxide nanomaterial—a review, Dev. Built Environ. 1 (2020), 100002.
[16] Y. Lin, H. Du, Graphene reinforced cement composites: a review, Construct. Build Mater. 265 (2020), 120312.
[17] L. Zhao, X. Guo, L. Song, Y. Song, G. Dai, J. Liu, An intensive review on the role of graphene oxide in cement-based materials, Construct. Build Mater. 241 (2020), 117939.
[18] N. Makul, Advanced smart concrete—a review of current progress, benefits and challenges, J. Clean. Prod. 274 (2020), 122899.
[19] W.J. Long, Y.C. Gu, H. Ma, H.D. Li, F. Xing, Mitigating the electromagnetic radiation by coupling use of waste cathode-ray tube glass and graphene oxide on cement composites, Compos, B 168 (2019) 25–33.
[20] L.L. Dong, W.G. Chen, N. Deng, C.H. Zheng, A novel fabrication of graphene by chemical reaction with a green reductant, Chem. Eng. J. 306 (2016) 754–762.
[21] H. Xiao, Y. Zongyou, W. Shixin, Q. Xiaoying, H. Qiyuan, Z. Qichun, Y. Qingyu, B. Freddy, Z. Hua, Graphene-based materials: synthesis, characterization, properties, and applications, Small 7 (14) (2011) 1876–1902.
[22] Z. Yanwu, M. Shanthi, C. Weiwei, L. Xuesong, S.J. Won, J.R. Potts, R.S. Ruoff, Graphene and graphene oxide: synthesis, properties, and applications, ChemInform 22 (35) (2010) 3906–3924.
[23] H. Chang, H. Wu, Graphene-based nanomaterials: synthesis, properties, and optical and optoelectronic applications, Adv. Funct. Mater. 23 (16) (2013) 1984–1997.

[24] S.H. Lv, L.J. Deng, W.Q. Yang, Q.F. Zhou, Y.Y. Cui, Fabrication of polycarboxylate/graphene oxide nanosheet composites by copolymerization for reinforcing and toughening cement composites, Cem. Concr. Compos. 66 (2016) 1–9.
[25] M. Han, Y. Muhammad, Y. Wei, Z. Zhu, J. Huang, J. Li, A review on the development and application of graphene based materials for the fabrication of modified asphalt and cement, Ceonstr. Build. Mater. 285 (2021), 122885.
[26] O.C. Compton, S.B.T. Nguyen, Graphene oxide, highly reduced graphene oxide, and graphene: versatile building blocks for carbon-based materials, Small 6 (6) (2010) 711–723.
[27] W. Meng, K.H. Khayat, Mechanical properties of ultra-high-performance concrete enhanced with graphite nanoplatelets and carbon nanofibers, Compos. B 107 (2016) 113–122.
[28] I. Rhee, Y.A. Kim, G.O. Shin, H.K. Ji, H. Muramatsu, Compressive strength sensitivity of cement mortar using rice husk-derived graphene with a high specific surface area, Construct. Build Mater. 96 (2015) 189–197.
[29] T.S. Qureshi, D.K. Panesar, Nano reinforced cement paste composite with functionalized graphene and pristine graphene nanoplatelets, Compos. B 197 (2020), 108-063.
[30] W. Ren, H.-M. Cheng, The global growth of graphene, Nat. Nanotechnol. 9 (2014) 726.
[31] A. Giménez-Pérez, S.K. Bikkarolla, J. Benson, C. Bengoa, F. Stüber, A. Fortuny, A. Fabregat, J. Font, P. Papakonstantinou, Synthesis of N-doped and non-doped partially oxidised graphene membranes supported over ceramic materials, J. Mater. Sci. 51 (18) (2016) 8346–8360.
[32] W.J. Long, J.J. Wei, H. Ma, F. Xing, Dynamic mechanical properties and microstructure of graphene oxide Nanosheets reinforced cement composites, Nanomaterials 7 (12) (2017) 407.
[33] L. Senff, D. Hotza, S. Lucas, V.M. Ferreira, J.A. Labrincha, Effect of nano-SiO2 and nano-TiO2 addition on the rheological behavior and the hardened properties of cement mortars, Mater. Sci. Eng. A 532 (2012) 354–361.
[34] D. Kong, X. Du, W. Su, Z. Hua, Y. Yang, S.P. Shah, Influence of nano-silica agglomeration on microstructure and properties of the hardened cement-based materials, Construct. Build Mater. 37 (3) (2012) 707–715.
[35] K. Al-Jabri, H. Shoukry, Use of nano-structured waste materials for improving mechanical, physical and structural properties of cement mortar, Construct. Build Mater. 73 (2014) 636–644.
[36] X. Li, A.H. Korayem, C. Li, Y. Liu, H. He, J.G. Sanjayan, W.H. Duan, Incorporation of graphene oxide and silica fume into cement paste: a study of dispersion and compressive strength, Construct. Build Mater. 123 (2016) 327–335.
[37] Y. Shang, D. Zhang, C. Yang, Y. Liu, Y. Liu, Effect of graphene oxide on the rheological properties of cement pastes, Construct. Build Mater. 96 (2015) 20–28.
[38] S. Chithra, S.R.R.S. Kumar, K. Chinnaraju, The effect of colloidal nano-silica on workability, mechanical and durability properties of high performance concrete with copper slag as partial fine aggregate, Construct. Build Mater. 113 (2016) 794–804.
[39] R. Laila, B. James, A. Rouhollah, M. Jon, S. Taijiro, Cement and concrete nanoscience and nanotechnology, Materials 3 (2) (2010) 918–942.
[40] H. Du, H.J. Gao, S.D. Pang, Improvement in concrete resistance against water and chloride ingress by adding graphene nanoplatelet, Cem. Concr. Res. 83 (2016) 114–123.
[41] F. Babak, H. Abolfazl, R. Alimorad, G. Parviz, Preparation and mechanical properties of graphene oxide: cement nanocomposites, Sci. World J. 4 (2014), 276323.

[42] H. Kesong, D.D. Kulkarni, I. Choi, V.V. Tsukruk, Graphene-polymer nanocomposites for structural and functional applications, Prog. Polym. Sci. 39 (11) (2014) 1934–1972.
[43] S. Parveen, S. Rana, R. Fangueiro, A review on nanomaterial dispersion, microstructure, and mechanical properties of carbon nanotube and nanofiber reinforced cementitious composites, J. Nanomater. 2013 (2013), 710175.
[44] W.J. Long, H.D. Li, C.L. Fang, F. Xing, Uniformly dispersed and re-agglomerated graphene oxide-based cement pastes: a comparison of rheological properties, mechanical properties and microstructure, Nanomaterials 8 (1) (2018) 31.
[45] C. Stephens, L. Brown, F. Sanchez, Quantification of the re-agglomeration of carbon nanofiber aqueous dispersion in cement pastes and effect on the early age flexural response, Carbon 107 (2016) 482–500.
[46] Z. Li, X. Guo, C. Ge, L. Qi, L. Guo, S. Xin, J. Liu, Investigation of the effectiveness of PC@GO on the reinforcement for cement composites, Construct. Build Mater. 113 (2016) 470–478.
[47] S. Lv, J. Zhang, L. Zhu, C. Jia, Preparation of cement composites with ordered microstructures via doping with graphene oxide nanosheets and an investigation of their strength and durability, Materials 9 (11) (2016) 924.
[48] L. Lu, D. Ouyang, Properties of cement mortar and ultra-high strength concrete incorporating graphene oxide nanosheets, nanomaterials, Nanomaterials 7 (7) (2017) 187.
[49] G. Bastos, F. Patiã, J. Armesto, Nano-inclusions applied in cement-matrix composites: a review, Materials 9 (12) (2016).
[50] K. Vallurupalli, W. Meng, J. Liu, K.H. Khayat, Effect of graphene oxide on rheology, hydration and strength development of cement paste, Construct. Build Mater. 265 (2020), 120311.
[51] L. Zhao, S. Zhu, H. Wu, X. Zhang, Q. Tao, L. Song, Y. Song, X. Guo, Deep research about the mechanisms of graphene oxide (GO) aggregation in alkaline cement pore solution, Construct. Build Mater. 247 (2020), 118446.
[52] X. Yan, D. Zheng, H. Yang, H. Cui, M. Monasterio, Y. Lo, Study of optimizing graphene oxide dispersion and properties of the resulting cement mortars, Construct. Build Mater. 257 (2020), 119477.
[53] J. Lin, E. Shamsaei, F. Basquiroto de Souza, K. Sagoe-Crentsil, W.H. Duan, Dispersion of graphene oxide–silica nanohybrids in alkaline environment for improving ordinary Portland cement composites, Cem. Concr. Compos. 106 (2020), 103488.
[54] W.J. Long, C. Fang, J. Wei, H. Li, Stability of GO modified by different dispersants in cement paste and its related mechanism, Materials 11 (5) (2018) 834.
[55] T.S. Qureshi, D.K. Panesar, Impact of graphene oxide and highly reduced graphene oxide on cement based composites, Construct. Build Mater. 206 (2019) 71–83.
[56] X. Li, C. Li, Y. Liu, J.C. Shu, C.M. Wang, J.G. Sanjayan, H.D. Wen, Improvement of mechanical properties by incorporating graphene oxide into cement mortar, Mech. Adv. Mater. Struct. (1) (2016).
[57] S. Chuah, W. Li, S.J. Chen, J.G. Sanjayan, W.H. Duan, Investigation on dispersion of graphene oxide in cement composite using different surfactant treatments, Construct. Build Mater. 161 (2018) 519–527.
[58] F. Sanchez, L. Zhang, Molecular dynamics modeling of the interface between surface functionalized graphitic structures and calcium-silicate-hydrate: interaction energies, structure, and dynamics, J. Colloid Interface Sci. 323 (2) (2008) 349–358.
[59] W. Min, R. Wang, Y. Hao, Z. Wang, S. Zheng, Adsorption characteristics of graphene oxide nanosheets on cement, RSC Adv. 6 (68) (2016). https://doi.org/10.1039.C6RA10902K.

[60] X. Li, Y.M. Liu, W.G. Li, C.Y. Li, J.G. Sanjayan, W.H. Duan, Z. Li, Effects of graphene oxide agglomerates on workability, hydration, microstructure and compressive strength of cement paste, Construct. Build Mater. 145 (2017) 402–410.
[61] X. Li, Z. Lu, S. Chuah, W. Li, Y. Liu, W.H. Duan, Z. Li, Effects of graphene oxide aggregates on hydration degree, sorptivity, and tensile splitting strength of cement paste, Compos. A 100 (2017) 1–8.
[62] W. Qin, X. Cui, W. Jian, S. Li, C. Lv, Y. Dong, Effect of fly ash on rheological properties of graphene oxide cement paste, Constr. Build. Mater. 138 (Complete) (2017) 35–44.
[63] W. Min, Y. Hao, R. Wang, S. Zheng, Chemically functionalized graphene oxide as the additive for cement–matrix composite with enhanced fluidity and toughness, Construct. Build Mater. 150 (2017) 150–156.
[64] C. Qiu, Q. Zhou, S. Lv, Y. Ma, Regulation of GO on cement hydration crystals and its toughening effect, Mag. Concr. Res. 65 (20) (2013) 1246–1254.
[65] M.M. Mokhtar, S.A. Abo-El-Enein, M.Y. Hassaan, M.S. Morsy, M.H. Khalil, Mechanical performance, pore structure and micro-structural characteristics of graphene oxide nano platelets reinforced cement, Construct. Build Mater. 138 (C) (2017) 333–339.
[66] W. Min, R. Wang, Y. Hao, S. Farhan, S. Zheng, C. Du, Study on the three dimensional mechanism of graphene oxide nanosheets modified cement, Construct. Build Mater. 126 (2016) 730–739.
[67] S. Lv, J. Liu, T. Sun, Y. Ma, Q. Zhou, Effect of GO nanosheets on shapes of cement hydration crystals and their formation process, Construct. Build Mater. 64 (64) (2014) 231–239.
[68] W. Li, X. Li, J.C. Shu, M.L. Yan, H.D. Wen, S.P. Shah, Effects of graphene oxide on early-age hydration and electrical resistivity of Portland cement paste, Constr. Build. Mater. 136 (Complete) (2017) 506–514.
[69] W. Qin, W. Jian, C.X. Lu, B.W. Liu, K. Zhang, C.Z. Li, Influence of graphene oxide additions on the microstructure and mechanical strength of cement, New Carbon Mater. 30 (4) (2015) 349–356.
[70] W.X. Bao, C.C. Zhu, W.Z. Cui, Simulation of Young's modulus of single-walled carbon nanotubes by molecular dynamics, Phys. B Condensed Matter. 352 (1–4) (2004) 156–163.
[71] W.J. Long, J.J. Wei, F. Xing, K.H. Khayat, Enhanced dynamic mechanical properties of cement paste modified with graphene oxide nanosheets and its reinforcing mechanism, Cem. Concr. Compos. 93 (2018) 127–139.
[72] M. Cao, H. Zhao, C. Zhang, Effect of graphene on mechanical properties of cement mortars, J. Cent. South Univ. 23 (4) (2016) 919–925.
[73] Z. Li, X. Guo, C. Ge, L. Qi, L. Guo, S. Xin, J. Liu, Mechanical behavior and toughening mechanism of polycarboxylate superplasticizer modified graphene oxide reinforced cement composites, Compos, B 113 (2017) 308–316.
[74] S. Lv, Y. Ma, C. Qiu, T. Sun, J. Liu, Q. Zhou, Effect of graphene oxide nanosheets of microstructure and mechanical properties of cement composites, Construct. Build Mater. 49 (12) (2013) 121–127.
[75] A.A. Gholampour, M.V. Kiamahalleh, D.N.H. Tran, T. Ozbakkaloglu, D. Losic, Revealing the dependence of the physiochemical and mechanical properties of cement composites on graphene oxide concentration, RSC Adv. 7 (87) (2017) 55148–55156.
[76] Q. Hao, W. Wei, H.H. Yun, Synergistic effect of graphene-oxide-doping and microwave-curing on mechanical strength of cement, J. Phys. Chem. Solid 103 (Complete) (2017) 67–72.
[77] S. Sharma, N.C. Kothiyal, Influence of graphene oxide as dispersed phase in cement mortar matrix in defining the crystal patterns of cement hydrates and its effect on mechanical, microstructural and crystallization properties, RSC Adv. 5 (65) (2015) 52642–52657.

[78] R.Y.N. Gengler, D.S. Badali, D. Zhang, K. Dimos, K. Spyrou, D. Gournis, R.J.D. Miller, Revealing the ultrafast process behind the photoreduction of graphene oxide, Nat. Commun. 4 (10) (2013) 2560.

[79] T. Chang, H. Gao, Size-dependent elastic properties of a single-walled carbon nanotube via a molecular mechanics model, J. Mech. Phys. Solids 51 (6) (2003) 1059–1074.

[80] T. Qureshi, D.K. Panesar, Impact of graphene oxide and highly reduced graphene oxide on cement based composite, Construct. Build Mater. 206 (2019) 71–83.

[81] B. Wang, R. Jiang, Z. Wu, Investigation of the mechanical properties and microstructure of graphene nanoplatelet-cement composite, Nanomaterials 6 (11) (2016) 200.

[82] W.J. Long, Y.C. Gu, B.X. Xiao, Q.M. Zhang, F. Xing, Micro-mechanical properties and multi-scaled pore structure of graphene oxide cement paste: synergistic application of nanoindentation, X-ray computed tomography, and SEM-EDS analysis, Construct. Build Mater. 179 (2018) 661–674.

[83] M.E. Abrishami, V. Zahabi, Reinforcing graphene oxide/cement composite with NH 2 functionalizing group, Bull. Mater. Sci. 39 (4) (2016) 1073–1078.

[84] H.K. Kim, I.S. Park, H.K. Lee, Improved piezoresistive sensitivity and stability of CNT/cement mortar composites with low water–binder ratio, Compos. Struct. 116 (1) (2014) 713–719.

[85] W.J. Long, J.J. Wei, Y.C. Gu, F. Xing, Research on dynamic mechanical properties of alkali activated slag concrete under temperature-loads coupling effects, Construct. Build Mater. 154 (2017) 687–696.

[86] S. Muthusamy, S. Wang, D.D.L. Chung, Unprecedented vibration damping with high values of loss modulus and loss tangent, exhibited by cement–matrix graphite network composite, Carbon 48 (5) (2010) 1457–1464.

[87] H. Cui, X. Yan, L. Tang, F. Xing, Possible pitfall in sample preparation for SEM analysis—a discussion of the paper "fabrication of polycarboxylate/graphene oxide nanosheet composites by copolymerization for reinforcing and toughening cement composites" by Lv et al, Cem. Concr. Compos. 77 (2017) 81–85.

[88] C. Lin, W. Wei, Y.H. Hu, Catalytic behavior of graphene oxide for cement hydration process, J. Phys. Chem. Solid 89 (2016) 128–133.

[89] S. Ghazizadeh, P. Duffour, N. Skipper, M. Billing, Y. Bai, An investigation into the colloidal stability of graphene oxide nano-layers in alite paste, Cem. Concr. Res. 99 (2017) 116–128.

[90] A. Nasser, A. Clément, S. Laurens, A. Castel, Influence of steel–concrete interface condition on galvanic corrosion currents in carbonated concrete, Corros. Sci. 52 (9) (2010) 2878–2890.

[91] W.J. Long, Y.C. Gu, F. Xing, K.H. Khayat, Microstructure development and mechanism of hardened cement paste incorporating graphene oxide during carbonation, Cem. Concr. Compos. 94 (2018) 72–84.

[92] A. Mohammed, J.G. Sanjayan, A. Nazari, N.T.K. Al-Saadi, The role of graphene oxide in limited long-term carbonation of cement-based matrix, Construct. Build Mater. 168 (2018) 858–866.

[93] A. Mohammed, J.G. Sanjayan, W.H. Duan, A. Nazari, Incorporating graphene oxide in cement composites: a study of transport properties, Construct. Build Mater. 84 (2015) 341–347.

[94] H. Du, S.D. Pang, Enhancement of barrier properties of cement mortar with graphene nanoplatelet, Cem. Concr. Res. 76 (2015) 10–19.

[95] L.C. Zhao, Preparation of graphene-cement paste anode for chloride extraction from marine reinforced concrete structures, Int. J. Electrochem. Sci. 11 (11) (2016) 9245–9253.

[96] M.Á. Climent, J. Carmona, P. Garcés, Graphite–cement paste: a new coating of reinforced concrete structural elements for the application of electrochemical anticorrosion treatments, Coatings 6 (3) (2016) 32.
[97] L. Zhao, D. Hou, P. Wang, X. Guo, Y. Zhang, J. Liu, J. Zhang, Experimental and molecular dynamics studies on the durability of sustainable cement-based composites: reinforced by graphene, Construct. Build Mater. 257 (2020), 119566.
[98] I. Arribas, I. Vegas, V. García, R.V.D.L. Villa, S. Martínez-Ramírez, M. Frías, The deterioration and environmental impact of binary cements containing thermally activated coal mining waste due to calcium leaching, J. Clean. Prod. 183 (2018) 887–897.
[99] J. Jain, N. Neithalath, Analysis of calcium leaching behavior of plain and modified cement pastes in pure water, Cem. Concr. Compos. 31 (3) (2009) 176–185.
[100] W.J. Long, T.H. Ye, Y.C. Gu, H.D. Li, F. Xing, Inhibited effect of graphene oxide on calcium leaching of cement pastes, Construct. Build Mater. 202 (2019) 177–188.
[101] W.J. Long, T.H. Ye, L.X. Li, G.L. Feng, Electrochemical characterization and inhibiting mechanism on calcium leaching of graphene oxide reinforced cement composites, Nanomaterials 9 (2) (2019) 288.
[102] W. Jiang, X. Li, L. Yang, M. Zhou, Z. Yu, Cement-based materials containing graphene oxide and polyvinyl alcohol fiber: mechanical properties, durability, and microstructure, Nanomaterials 8 (9) (2018) 638.
[103] A. Mohammed, J.G. Sanjayan, W.H. Duan, A. Nazari, Graphene oxide impact on hardened cement expressed in enhanced freeze–thaw resistance, J. Mater. Civ. Eng. 28 (9) (2016), 04016072.
[104] Y. Mostafa, R.J.-M. Pellenq, Y. Bilge, Glassy nature of water in an ultraconfining disordered material: the case of calcium-silicate-hydrate, J. Am. Chem. Soc. 133 (8) (2011) 2499–2510.
[105] T. Tong, F. Zhou, Q. Liu, S. Wang, S. Tan, Q. Yu, Investigation of the effects of graphene and graphene oxide nanoplatelets on the micro- and macro-properties of cementitious materials, Construct. Build Mater. 106 (2016) 102–114.
[106] H. Du, S.T. Quek, S.D. Pang, Smart multifunctional cement mortar containing graphite nanoplatelet, in: Proceedings Volume 8692, Sensors and Smart Structures Technologies for Civil, Mechanical, and Aerospace Systems, San Diego, United States, 2013.
[107] S.D. Pang, H.J. Gao, C. Xu, S.T. Quek, H. Du, Strain and damage self-sensing cement composites with conductive graphene nanoplatelet, in: Proceedings Volume 9061, Sensors and Smart Structures Technologies for Civil, Mechanical, and Aerospace Systems, San Diego, United States, 2014.
[108] A. Al-Dahawi, O. Öztürk, F. Emami, G. Yıldırım, M. Şahmaran, Effect of mixing methods on the electrical properties of cementitious composites incorporating different carbon-based materials, Construct. Build Mater. 104 (2016) 160–168.
[109] M. Saafi, L. Tang, J. Fung, M. Rahman, J. Liggat, Enhanced properties of graphene/fly ash geopolymeric composite cement, Cem. Concr. Res. 67 (67) (2015) 292–299.
[110] J. Ming, L. Jiang, M. Lu, S. Bai, Monitoring chloride ion penetration in concrete structure based on the conductivity of graphene/cement composite, Constr. Build. Mater. 136 (Complete) (2017) 394–404.
[111] S. Sun, B. Han, J. Shan, Y. Xun, Y. Wang, H. Li, J. Ou, Nano graphite platelets-enabled piezoresistive cementitious composites for structural health monitoring, Constr. Build. Mater. 136 (Complete) (2017) 314–328.
[112] S. Sun, S. Ding, B. Han, S. Dong, Y. Xun, D. Zhou, J. Ou, Multi-layer graphene-engineered cementitious composites with multifunctionality/intelligence, Compos. B 129 (2017) 221–232.
[113] I. Rhee, J.S. Lee, Y.A. Kim, H.K. Jin, H.K. Ji, Electrically conductive cement mortar: incorporating rice husk-derived high-surface-area graphene, Construct. Build Mater. 125 (2016) 632–642.

[114] A. Al-Dahawi, M.H. Sarwary, O. Öztürk, G. Yıldırım, A. Akın, M. Şahmaran, M. Lachemi, Electrical percolation threshold of cementitious composites possessing self-sensing functionality incorporating different carbon-based materials, Smart Mater. Struct. 25 (10) (2016), 105005.

[115] Z. Wang, B. Zhang, D. Guan, Take responsibility for electronic-waste disposal, Narture 536 (7614) (2016) 23–25.

[116] J.X. Lu, S.P. Chi, Improvement of early-age properties for glass-cement mortar by adding nanosilica, Cem. Concr. Compos. 89 (2018) 18–30.

[117] B. Zhang, H. Tan, W. Shen, G. Xu, B. Ma, X. Ji, Nano-silica and silica fume modified cement mortar used as surface protection material to enhance the impermeability, Cem. Concr. Compos. (2018). S0958946518300283.

[118] W.J. Long, Y.C. Gu, F. Xing, K.H. Khayat, Evaluation of the inhibiting effect of graphene oxide on lead leaching from waste cathode-ray tube glass incorporated in cement mortar, Cem. Concr. Compos. 104 (2019).

[119] W.J. Long, Y.C. Gu, D. Zheng, N. Han, Utilization of graphene oxide for improving the environmental compatibility of cement-based materials containing waste cathode-ray tube glass, J. Clean. Prod. 192 (2018) 151–158.

[120] C. Zongping, X. Chuan, M. Chaoqun, R. Wencai, C. Hui-Ming, Lightweight and flexible graphene foam composites for high-performance electromagnetic interference shielding, Adv. Mater. 25 (9) (2013) 1296–1300.

[121] D.X. Yan, H. Pang, L. Bo, R. Vajtai, X. Ling, P.G. Ren, J.H. Wang, Z.M. Li, Structured reduced graphene oxide/polymer composites for ultra-efficient electromagnetic interference shielding, Adv. Funct. Mater. 25 (4) (2015) 559–566.

[122] A.H. Frey, Headaches from cellular telephones: are they real and what are the implications? Environ. Health Perspect. 106 (3) (1998) 101–103.

[123] S. Avanish Pratap, M. Monika, C. Amita, S.K. Dhawan, Graphene oxide/ferrofluid/cement composites for electromagnetic interference shielding application, Nanotechnology 22 (46) (2011), 465701.

[124] Z. Xu, C. Gao, Aqueous liquid crystals of graphene oxide, ACS Nano 5 (4) (2011) 2908–2915.

[125] A.R. Alian, M.A.N. Dewapriya, S.A. Meguid, Molecular dynamics study of the reinforcement effect of graphene in multilayered polymer nanocomposites, Mater. Des. 124 (2017) 47–57.

[126] A. Sedaghat, M.K. Ram, A. Zayed, R. Kamal, N. Shanahan, Investigation of physical properties of graphene-cement composite for structural applications, J. Thermoplast. Compos. Mater. 4 (4) (2014) 12–21.

[127] C. Xia, S. Sun, B. Han, Y. Xun, J. Ou, Thermal and electromagnetic properties of nanographite platelets modified cementitious composites, Compos. A 93 (2016) 49–58.

[128] H. Chu, J. Jiang, W. Sun, M. Zhang, Effects of graphene sulfonate nanosheets on mechanical and thermal properties of sacrificial concrete during high temperature exposure, Cem. Concr. Compos. 82 (2017).

[129] G. Jing, Z. Ye, J. Wu, S. Wang, X. Cheng, V. Strokova, V. Nelyubova, Introducing reduced graphene oxide to enhance the thermal properties of cement composites, Cem. Concr. Compos. 109 (2020), 103559.

[130] Z. Lu, J. Yu, J. Yao, D. Hou, Experimental and molecular modeling of polyethylene fiber/cement interface strengthened by graphene oxide, Cem. Concr. Compos. 112 (2020), 103676.

CHAPTER 7

Use of nanomaterials in geopolymers

Nima Farzadnia[a] **and Caijun Shi**[b,c]
[a]Department of Civil, Geological, and Environmental Engineering, University of Alaska Fairbanks, Fairbanks, AK, United States
[b]Key Laboratory for Green & Advanced Civil Engineering Materials and Application Technology of Hunan Province, International Innovation Center for Green & Advanced Civil Engineering Materials of Hunan Province, College of Civil Engineering, Hunan University, Changsha, China
[c]Department of Civil Engineering, The University of British Columbia, Vancouver, BC, Canada

7.1 Introduction

Nanomaterials have received significant attention because of their unique functional properties. They have a great surface area to volume ratio, which can lead to greater chemical reactivity. This property has the potential for greater impacts in various fields. For example, using nanomaterials in construction materials can lead to generating products with advanced functions. Generally, nanomaterials can reduce the amount of cement used and improve concrete's mechanical and physical properties. Also, coating construction materials with nanoparticles can provide better water repulsion and corrosion resistance, extending their life service. Alkali-activated materials (or geopolymer), as an alternative to conventional concrete, can also be benefited from nanotechnology.

Geopolymer concrete is of interest in recent years due to its desirable properties, such as comparable mechanical properties to conventional concrete and, in some cases, higher durability owing to its low permeability, resistance to acids, and high frost durability [1]. The geopolymer binders are easier to produce and less energy-intensive than Portland cement. It is mainly because their production does not require a high temperature for the calcination of raw materials. Such characteristics can make geopolymers promising building materials for general construction and toxic waste immobilizing, sealing, and temperature-resistant applications [2]. Nanomaterials can improve the mechanical and physical properties of geopolymers in different fashions. This chapter discusses the effects of the widely used nanomaterials on reaction kinetics, chemical characterization of reaction products, pore structure, mechanical properties, as well as durability of

Nanotechnology for Civil Infrastructure
https://doi.org/10.1016/B978-0-12-817832-4.00003-1

geopolymers. In this chapter, the term "geopolymer" is used as an interchangeable term for "alkali-activated materials."

7.2 Geopolymers

7.2.1 Definitions and classification

Geopolymers were first noticed by Kuhl [3] in the early 1900. He reported the activation of an aluminosilicate (slag) with an alkali. In the 1950s, Glukhovsky [4], for the first time, developed binders by activating low-calcium aluminosilicate precursors using alkali metals. He termed the binders "soil cements." In the 1970s, Davidovits [5] developed an alkali-activated binder based on metakaolin (MK) for high fire-resistant polymer applications. The growing trend to develop "alkali-activated materials" for the production of construction materials started later in the 1980s [6].

The "alkali activation" is the generic term in geopolymers that refers to the activation of a solid aluminosilicate (termed the "precursor") under alkaline conditions (induced by the "alkali activator") to produce a binding gel that hardens over time. These binders are generated from the alkali activation of a wide range of aluminosilicates using alkalis such as sodium hydroxide, sodium silicates, and sodium carbonates. Shi et al. [1] divided alkali-activated materials (AAMs) based on the precursors into five categories: (1) alkali-activated slag-based cements; (2) alkali-activated Portland blended cements; (3) alkali-activated pozzolan cements, (4) alkali-activated lime-pozzolan/slag cements, and (5) alkali-activated calcium aluminate blended cement. Provis [7] divided them into two classifications: low-calcium alkali-activated binders and high-calcium alkali-activated binders. Due to the diversity of the precursors used in the development of the geopolymers, they are locally adaptable. They can, hence, form a key component in the development of future sustainable construction materials. Geopolymers are considered sustainable alternatives for ordinary Portland cement (OPC) [6,8].

The formation of geopolymer can be classified into three stages [9]. The first stage is called "destruction-coagulation." The Si—O—Si, Al—O—Al, and Al—O—Si bonds disaggregate from the precursor. The alkaline metals act as electron donor atoms and redistribute the electronic density around the silicon atoms. They also hinder reaction reversibility after polymerization by generating Si—O—Na^+ bonds, which contribute to the development of the coagulated structure through the transport of the reacting structural units (stage 2). At this stage, "also known as coagulation-condensation," transport and accumulation of the reacting structural unit enhance contact among the disaggregated products leading to polycondensation of disaggregated bonds. The third stage is the "condensation-crystallization," in which the particles from the precursor and the condensation outcome form the activation products. The mineralogical composition of the precursor and the type and content of the activator determine the polymerization products. The primary reaction product in low calcium-bearing materials is an alkali aluminosilicate hydrate (N—A—S—H) gel. Using a high-calcium-bearing material such as slag leads to the formation of calcium aluminosilicate hydrate (C—A—S—H) gel. The former gel, N—A—S—H, is poor in calcium and is more durable due to the lower content of the Ca ions. The coexistence of N—A—S—H and C—A—S—H has been beneficial in the geopolymer matrix [10].

7.2.2 Geopolymer constituents: Precursors, activators, and admixtures

Geopolymers usually consist of two components: cementitious components and alkaline activators. In some cases, the third component as an admixture is added to comply with the requirement of the application. There have been extensive studies and literature reviews on the effect of constituents in geopolymers [6]. A wide array of precursors, namely, industrial by-products, agriculture waste, and several aluminosilicate raw materials, have been used to produce geopolymers. Most of them are the same as what is used as supplementary cementitious materials (SCMs) in blends with Portland cement. However, other sources available in the form of ashes can be employed in the production of geopolymers. Such sources can be clays, lateritic, volcanic soils, and agricultural wastes. Usage of precursors in geopolymers depends on the properties expected and influential factors such as the region and the type of locally available materials. The dominant characteristics of precursors are their content of active CaO, Al_2O_3, and SiO_2 and their particle size which increase their reactivity.

Chemically activating the precursors, caustic alkalis or alkaline salts are prominently used as alkaline activators in geopolymers. The primary source of alkalinity in an activator is alkali metals such as sodium in sodium hydroxide, sodium silicates (also known as water glass), and sodium carbonates. The use of each activator can be largely dependent on the resources. For example, higher molarity of hydroxide activators is needed to activate precursors, especially with low calcium-bearing precursors [11], and even requires thermal curing (60°C higher), depending on the reactivity of the material [12].

This can cause difficulties for on-site casting. Therefore, adjustments of activators are also needed to tailor the geopolymer based on the required specification of the construction.

The third class of components that can be added to geopolymers are admixtures/additives. Such materials are added to enhance their workability, mechanical properties, stability, and durability. For example, the incorporation of sucrose and citric acid was reported as set-modifying agents [13]. They reported that sucrose retarded the polymerization because it generated insoluble metal complexes with Ca, Al, and Fe ions. On the other hand, citric acid acted as an accelerator and reduced the setting time at high dosage, unlike its role as a retarder in OPC systems at low dosage. Besides, the retarding effect of polycarboxylate-based superplasticizer was observed in activated slag/FA-based systems. The retardation of the setting improved the workability of the geopolymer [14]. The performance of other admixtures was also studied to add the functionality of the geopolymers. They included shrinkage-reducing agents [15], preheated materials, and thermomagnetic admixtures [16], as well as fibers such as basalt and polypropylene fibers [17]. This paper mainly focuses on the effect of nanomaterials on geopolymers.

7.2.3 Key engineering properties

Geopolymers' key engineering properties mainly include fresh properties, such as workability and rheology, and hardened properties. The hardened properties are compressive strength, elastic modulus, tensile strength, fracture performance, and durability. The hardened properties are a function of the reaction kinetics which will be explained in Section 6.3.1. The testing procedures for assessing the properties of geopolymers are similar to those for OPC systems. However, no specific testing standards are available exclusively for geopolymers. The engineering properties of geopolymers are studied with respect to their water-to-solid ratio, activator type and silicate modulus, water/Na_2O, curing time, and curing temperature. The fresh properties (e.g., slump) vary depending on the molarity and type of activator [18,19] and rheology modifying admixtures or extra water that are added to the mixture [20]. The setting time of geopolymer concrete is reported up to 120 min [21].

Several studies investigated the mechanical properties of geopolymers [22]. For example, it is reported that the splitting tensile strength of FA-based geopolymer was higher than that of OPC concrete. However, the modulus of elasticity of geopolymer was found to be lower than that of OPC concrete [23]. For the durability of geopolymers, extensive research has been done to investigate chemical or microbial corrosion, frost damage, surface abrasion, shrinkage, and carbonation of concrete [24]. Geopolymer concrete's durability was reported to correspond to factors such as properties and proportion of raw materials, alkaline solution type, molarity, and curing regime. Existing literature shows that the geopolymer has better performance than OPC concrete in terms of durability [25]. An extensive review has been conducted in [22,24], and further discussion is out of the scope of this chapter.

7.3 Effect of nanomaterials on setting, hardening, and microstructure of geopolymers

The properties of geopolymers are influenced by their microstructure development upon activation. Three mechanisms govern the microstructure development of geopolymers: (1) dissolution of aluminosilicate particles, (2) nucleation and growth of initial solid phases, and (3) interactions of formed phases at the boundaries [26,27]. Such interactions will continue until chemical equilibria in the pore solution are reached. Upon rendering solid phases onto the particles, diffusion of active ions through the reaction products will be the dominant mechanism. Such polymerization mechanisms and the products of geopolymerization are critical factors for the fresh and hardened properties of geopolymers.

As mentioned in Section 6.2.1, the main reaction product of high-calcium precursors is the C—A—S—H gel [28], with a similar structure to a disordered tobermorite, which is highly heterogeneous. However, the Ca/Si of the C—A—S—H gel depends on the activator type. Reportedly, the C—A—S—H in NaOH-activated slag is denser (higher Ca/Si) and more well ordered than the C—A—S—H type gel generated in silicate-activated binders [29]. For low-calcium precursors, such as MK and fly ash (FA) systems, the main product of the geopolymerized systems is N—A—S—H gel. It is a disordered and highly cross-linked aluminosilicate gel, with Si and Al tetrahedral coordination [30,31]. However, similar to high-calcium precursors, the fundamental binder structure depends highly on the activator. This section reviews the influence of nanosilica (NS), nanoclay, nanoalumina, and nanotitania on the reaction kinetics and products and the fresh and hardened properties of geopolymers.

7.3.1 Reaction kinetics

7.3.1.1 Nanosilica

NS is the most widely used nanomaterials in geopolymers. Silica nanoparticles can accelerate the polymerization process due to high surface energy, high reactivity, and chemical instability [32]. Previous studies [33,34] reported that the addition of NS increased the soluble silica content, which enhanced the dissolution rate of silica and alumina phases in the precursor, and promoted the formation of long-chain silicate oligomers. Adak et al. [35] investigated such an effect on FA-based geopolymers and reported a more rapid dissolution of Si and Al phases. This phenomenon accelerated the rate of polymerization. The authors showed that the incorporation of nano-SiO_2 increased the intensity bands of the Fourier-transform infrared spectroscopy (FTIR) spectrum related to the Si—O—T asymmetric vibration in the nanomodified geopolymers. This band determines the degree of polymerization [36]. Gao et al. [36] showed the high intensity of Si—O—T asymmetric vibration transmittance in the FTIR spectrum of the MK-based geopolymer, indicating a higher degree of polymerization. In general, the SiO_2/Na_2O of 1.5 with 1% NS obtained the highest degree of polymerization and led to the densification of the matrix. Tsai et al. [37] also reported a chemical shift of the fully polymerized silicon in MK-based geopolymer with NS. The authors showed the dominance of higher cross-links of Al in the system. Their results also elaborated that the Si atom environment changed, and the intensity of the main associated peaks increased with the incorporation of NS. This can indicate more formation of geopolymer products with the presence of nano-SiO_2. Similar findings were reported by FTIR analysis on the spent catalyst MK-based geopolymer with up to 2% NS by Lo et al. [38]. The authors found that peaks attributable to the geopolymers' Si—O—T symmetrical vibrations differed with variation in the NS content. The increase in the NS content shifted the band to lower frequencies. Different researchers also studied the effect of NS on the reaction kinetics of high-calcium precursors. Wang et al. [39] studied the effect of NS on the hydration of alkali-activated slag. The results showed that the total heat of hydration increased as NS content increased to 3% (Fig. 7.1). They explained that NS accelerated the hydration reaction in the early stages through Its nucleation effect. Also, they mentioned that a large specific surface and high reactivity of NS that was neutralized with alkali solution could release some heat.

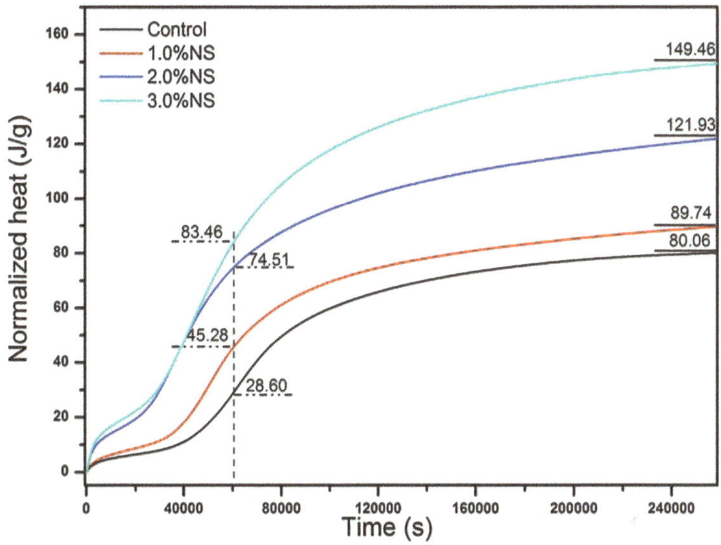

Fig. 7.1 Total hydration heat release curves of alkali-activated slag with NS [39].

7.3.1.2 Nanoclay

Nanoclay is another widely used nanomaterial extensively studied due to its high surface area to volume ratio and high chemical reactivity, promoting the pozzolanic reaction [40,41]. However, limited studies investigated the effects of nanoclay particles on the reaction kinetics of geopolymers. Like NS, nanoclay can enhance the polymerization reaction by increasing the silica content, leading to a higher matrix density. The enhanced reaction of FA-based geopolymer was reported by Assaedi et al. [42]. The FTIR results showed increased intensity and the area of stretching peaks associated with the Si—O—Si in geopolymer with nanoclay. This indicates a higher formation of the amorphous material and geopolymer gel. However, the higher content of nanoclay (more than 2%) remained inactive with no further improvement in polymerization.

7.3.1.3 Nanoalumina

Despite its high crystallinity, nanoalumina was reported to change the reaction kinetics of the geopolymers. However, more extensive research is needed to elaborate on the underlying mechanisms. Previous work showed that nanoalumina lowered the Si/Al due to their high surface area, resulting in more geopolymer gel formation [43,44]. It was also reported that the crystalline phase of nanoalumina did not participate in the polymerization reaction and may only act as nanofillers [45]. However, Hajimohammadi et al. [46] observed changes in the reaction kinetics of low-calcium geopolymer due to the nucleation effect of nanoparticles. Nanoalumina particles acted as seed particles and accelerated the formation of geopolymer gel. This enabled a higher rate of growth and accumulation of N—A—S—H gel [47]. Also, Alomayri [48] reported that nanoalumina acted as a nanofiller and densified the FA-based geopolymer. They showed that nanoalumina effectively strengthened the interfacial bonding of the matrix, which ultimately increased the hardened properties of the geopolymer.

7.3.1.4 Nanotitania

Among nanomaterials with crystal form, nanotitania is generally used for its photocatalytic properties. However, the enhancing effect of nanotitania on the polymerization reaction of geopolymers has also been reported. Similar to nanoalumina, the crystallinity of nanotitania can provide nucleation sites for the ionic species in the pore solution, increasing the rate of polymerization. Duan et al. [49] showed an increased intensity of the crystal structure and higher formation of hydrates in FA-based geopolymer. The authors activated fluidized bed FA-based geopolymer using sodium silicate and sodium hydroxide with an alkali activator modulus of 1.5. Nanotitania was P25 (75% anatase and 25% rutile) with particle size ranging from 1 nm to 30 nm at addition levels of 1%, 3%, and 5%. The authors related the enhancing effect of nanotitania on the reaction kinetics to their hydration accelerating effect and their performance as nucleation sites in the geopolymer system. The results were indicative that the size and amount of unreacted particles decreased as the content of nanotitania elevated [49].

7.3.2 Characterization of reaction products

7.3.2.1 Nanosilica

NS mainly enhances the density of the matrix and improves the mechanical properties by increasing the formation of geopolymer products. Adak et al. [35] reported that the addition of NS to the low-calcium FA-based geopolymer rendered a new amorphous phase and led to the formation of crystalline phases in the matrix. They used NS in the colloidal form with a size of 4–16 nm and solid content of 31% in the colloidal solution. The authors reported that a large number of crystalline compounds, such as new phases of quartz (SiO_2), albite ($NaAlSi_3O_8$), kaolinite ($Al_2Si_2O_5(OH)_4$), alite (Ca_3SiO_5), mullite ($3Al_2O_3$-$2SiO_2$), and $Ca(OH)_2$ crystalline compounds were detected. Such crystalline products resulted from the transformation of the amorphous phase of FA in NS-modified geopolymer. The render of such crystalline compounds increased the mechanical properties of the system and enabled its production at room ambient temperature. The study by Gao et al. [50] also showed the higher formation of geopolymer products with the presence of nano-SiO_2 in MK-based geopolymer. The reaction product was an amorphous hydrated aluminosilicate. The TG/DTA results showed that NS increased the amount of structural water and gels, which benefited the structural development.

For high-calcium precursors, Phoo-ngernkham et al. [51] investigated the effect of NS on Class C FA-based geopolymer paste. They reported a higher formation of C—S—H, an increase in the density of quartz, and secondary products in the geopolymer matrix when NS was added. Another XRD investigation by [39] on alkali-activated slag with NS showed increased peak intensity of hydrated calcium aluminosilicate and hydrated calcium silicate gels. However, a study by Behfarnia and Rostami [52] showed a possible negative impact of NS on the microstructure of alkali-activated slag. They studied the effect of NS (20–30 nm) on alkali-activated slag at contents of 0.5%, 1%, 3%, and 5% by weight of slag. Three main phases, including C—A—S—H gel, calcium carbonate (calcite), and magnesium-aluminum carbonate hydroxide hydrate ($Mg_6Al_2CO_3(OH)_{16}4H_2O$), were detected. However, the results of the XRD analysis showed that the addition of NS formed a new calcium silicate carbonate phase with the mineral name Tilleyite. This phase has a layered and laminate structure [53] (Fig. 7.2). The authors reported that the formation of this

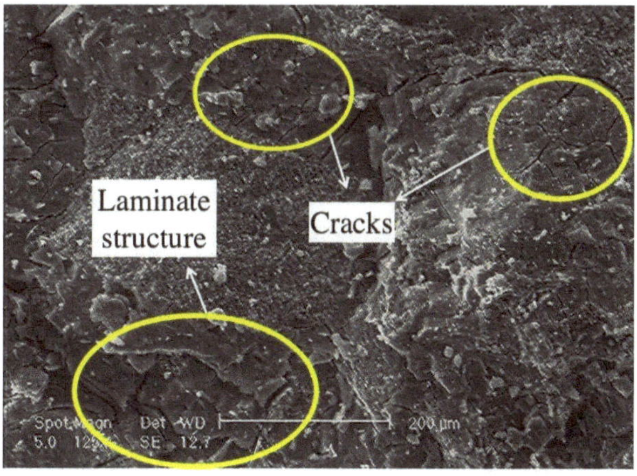

Fig. 7.2 SEM image of alkali-activated slag at 90 day with 1% NS [52].

new phase and its layered structure can lower the durability of activated slag due to the increased permeability.

7.3.2.2 Nanoclay

In a study by Assaedi et al. [42], low-calcium FA with a specific gravity of 2.1 was activated using a combination of sodium hydroxide solution and sodium silicate. The alkaline solution to FA was 0.75, and the ratio of sodium silicate solution to sodium hydroxide solution was 2.5. The nanoclay was natural montmorillonite modified with a quaternary ammonium salt particle with a particle size of 90% volume less than 13 nm. It was added to the FA at 1%, 2%, and 3% contents by weight. The XRD-and FTIR analysis showed that the addition of nanoclay increased the amorphous phase, which led to a denser matrix, and a lower content of unreacted fly ash particles. The XRD pattern of geopolymer with nanoclay showed an amorphous aluminosilicate phase was developed. FTIR results showed decreased peaks associated with physically free water and chemically bounded water through the hydrogen bond, which agrees with their results from the thermal analysis. They reported that the inclusion of nanoclay increased the content of the amorphous phases and polymerization process compared to the control geopolymer. In another study by Assaedi et al. [54], they reported similar results, although the formation of sodium carbonates was also detected. The authors showed that the nanoclay slowed down the reaction progress over time and reacted with more O—H groups (physical water), which led to the formation of a more robust material.

7.3.2.3 Nanoalumina

It is reported that nanoalumina has no significant influence on the polymerization process. However, the pore-filling effect of nanoalumina can improve the microstructure of geopolymer. Results from Alomayri [48] showed that the XRD pattern of FA-based geopolymer with nanoalumina (50 nm) was similar to those without nanoalumina. They concluded that nanoalumina did not participate in the geopolymeric reactions but existed as filler particles in the geopolymer matrix. However, Phoo-ngernkham et al. [51] reported similar behavior of nanoalumina to that of NS despite their distinct crystallinity. In their study, the XRD pattern of nanoalumina-modified FA-based geopolymer showed a higher formation of N—A—S—H gel in the presence of nanoalumina. Huang and Han [44] also confirmed the effectiveness of nanoalumina particles on polymerization. They reported that incorporating 5% nanoalumina shifted and intensified the Si—O—T (T refers to Si or Al) asymmetric stretching vibration. This indicates enhanced polymerization and the formation of higher geopolymer gel.

7.3.2.4 Nanotitania

A study by Duan et al. [49] on fluidized FA-based geopolymer with nanotitania showed that nano-TiO_2 particles improved the formation of hydrated structures, enhanced the crystalline structure, and increased the crystalline phase of albite. However, in a study by Guzmán-Aponte et al. [55], it was reported that the addition of up to 10% nano-TiO_2 to the MK-based geopolymer showed no significant change in the FTIR spectra compared to the reference mixture. Results from the X-ray diffraction and FTIR analyses of Llano-Guerrero et al. [56] also showed no new phases in the presence of the nanoparticles. The authors investigated the effect of nano-TiO_2 on MK-based geopolymer with the alkaline solution with mixtures of sodium silicate of a modulus of 3.25 and sodium hydroxide. The nanotitania was P-25 with a ratio Anatase/Rutile 80:20 and an average particle size of 21 nm. However, Yang et al. [57] showed a significant effect of nanotitania (20–100 nm) on alkali-activated slag. The results showed that the transmittance related to O—H stretching and molecular water changed noticeably. Nanotitania showed an increase in those bands, indicating a higher amount of chemically bound water and the formation of more hydration products. Also, bands associated with C—S—H/C—A—S—H gel were detected to grow wider when TiO_2 was added. Results were conclusive that more hydration products like C—S—H and C—A—S—H were produced.

7.3.3 Pore structure

7.3.3.1 Nanosilica

Previously, it was discussed that NS mainly induces the higher formation rate of geopolymer products, increasing the matrix's density and improving the mechanical properties. In a study by Deb et al. [58] on FA-based geopolymer, it was shown that the NS (15 nm) refined the pore structure through the filling effect and developed a more compact microstructure. They also showed that NS developed a denser microstructure by reducing the amount of unreacted FA particles compared to control samples. Gao et al. [50] also showed that the macropores in MK-based geopolymer transformed to mesopores with NS. This was mainly attributed to the enhanced polycondensation of the hydrated gels by the accelerating effect of NS on the dissolution of the precursor. It was reported that pores larger than 200 nm were reduced by a higher degree of activation in NS-containing geopolymer. Such pores represent the spaces between partially reacted or unreacted raw material and the geopolymer gel. Assaedi et al. [59] also reported that adding 3% NS helped further activation of FA particles to form geopolymer gel. They observed that NS acted more efficiently as a filler when mixing was performed in the dry conditions compared to wet conditions. However, Lo et al. [38] showed a threshold for NS usage in MK-based geopolymer. The use of more than 5% NS was not beneficial, as the higher NS content caused poor dispersion and lowered the enhancing influence of nanoparticles.

As for precursors with high-calcium content, Khater [60] studied the effect of NS on the microstructure of alkali-activated slag. The authors reported that the morphological texture of the matrix improved with the inclusion of NS which was mainly due to the filling of microvoids and the accelerated hydration reactions. NS particles also provided more nucleation sites for the growth of geopolymer gel. However, Wang et al. [39] reported the early-age effect of NS on the hydration process. They stated that NS could not act as a filler due to the higher reactivity of particles involved in the activation process. All studies agreed that agglomeration could happen when higher contents of NS are applied. Agglomeration of particles caused the formation of voids and led to a less homogeneous microstructure. Fig. 7.3 shows the reduced beneficial effect of NS in decreasing the porosity of MK-based geopolymer.

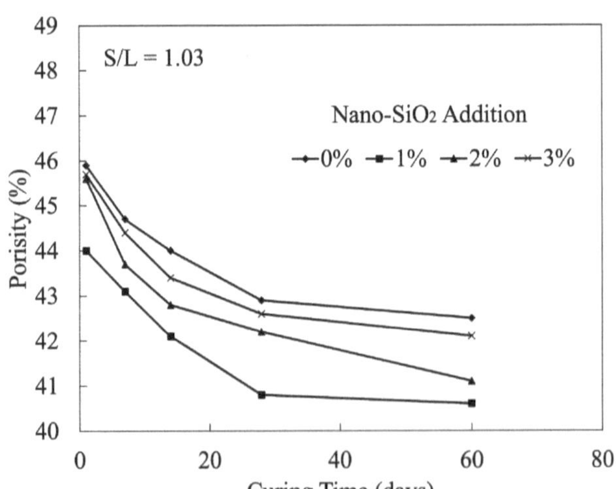

Fig. 7.3 Porosity of MK-based geopolymer with up to 3% NS [50].

7.3.3.2 Nanoclay

Assaedi et al. [42] studied the effect of using 3% nanoclay on the microstructure of FA-based geopolymer. SEM images showed a high concentration of unreacted and partially reacted FA particles in the plain geopolymer matrix. This increased the system's porosity; however, nanoclay refined the microstructure by enhancing the polymerization. Like NS, excessive content of nanoclay led to the agglomeration of nanoparticles and reduced the beneficial effect of nanoclay. Fig. 7.4 shows the agglomerated nanoclay in the matrix.

7.3.3.3 Nanoalumina

A few available studies investigated the microstructure of geopolymers with nanoalumina. However, it is reported that incorporating nanoalumina changes the Si/Al and influences the polymerization. Nanoalumina can enhance the microstructure of the geopolymer by filling nano-sized pores

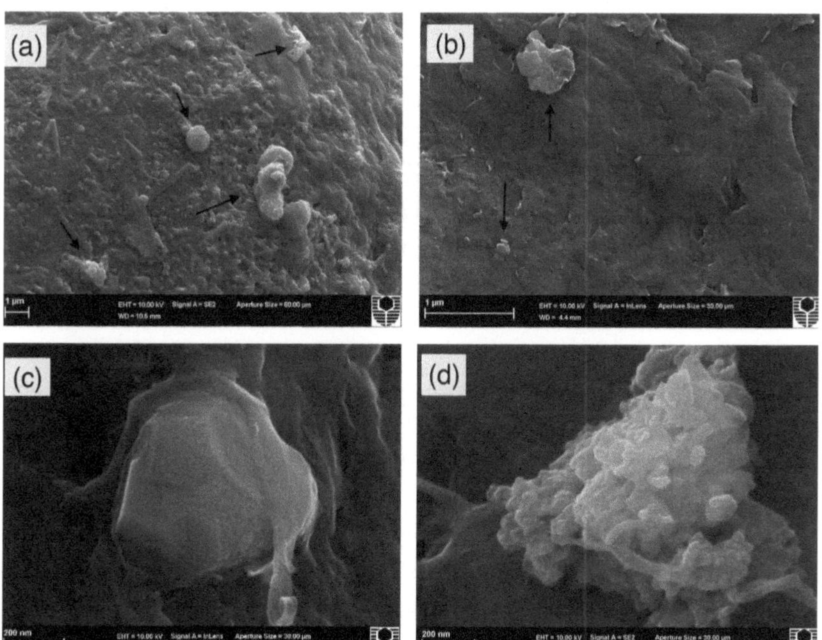

Fig. 7.4 SEM images of the fracture surface of geopolymer with different loadings of nanoclay with agglomerated nanoclay particles embedded in the matrix at (A and B) low magnification and (C and D) high modification [42].

as well as inducing chemical reactions. Huang and Han [44] investigated the use of nanoalumina with a high crystalline structure (α-Al_2O_3) in FA-based geopolymer. The authors reported that the aluminum phase of Al_2O_3 took part in geopolymeric reactions and resulted in a denser microstructure. Results from Alomayri [48] on the microstructure of FA-based geopolymer with nanoalumina also showed that the amount of nonreacted FA particles was reduced, and pores were significantly refined when 2% nanoalumina was added. The authors stated that nanoalumina contributed to the geopolymer reaction and acted as filler. However, the effectiveness of nanoalumina at higher content (3%) was reduced.

As for precursors with higher calcium, Phoo-ngernkham et al. [51] showed that nanoalumina had a very similar effect with NS on high-calcium FA-based geopolymer. They reported that nanoalumina resulted in the additional formation of C—S—H/C—A—S—H gels, which coexisted with N—A—S—H gel [60] and thus led to the overall densification of the matrix. The SEM images showed that the microstructure of geopolymer was enhanced when up to 2% nanoalumina was used. However, the excessive content of nanoparticles (e.g., 3%) resulted in the agglomeration of nanoparticles. Also, a negative impact on the microstructure was observed. Chindaprasirt et al. [61] also reported that nanoalumina positively affected the microstructure of high-calcium FA-geopolymer. Their study showed the existence of unreacted FA particles enveloped by C—S—H/C—S—A—H and N—A—S—H gel in control geopolymer samples. In contrast, the C—S—H/C—A—S—H gel was more evenly distributed in the matrix with nanoalumina, which resulted in a more homogeneous microstructure.

7.3.3.4 Nanotitania

Duan et al. [49] reported that nanotitania accelerated the hydration through the nucleation and filling effect. Fig. 7.5 shows the effects of using 1 to 5% TiO_2 nanoparticles on the pore size distribution and porosity of geopolymer at 28 days. The results showed that the pores of geopolymer were refined as the addition of nanotitania increased to 5%. The nanotitania addition shifted the most probable peaks of the pore size distribution to smaller sizes (Fig. 7.5A). Also, the hydrated products increased with increasing content of nanotitania leading to a decreased porosity and much denser microstructure related to the nucleation and filling effect of nano-TiO_2 (Fig. 7.5B). Maiti et al. [62] studied the effect of nano-TiO_2 particle size on the microstructure of FA-based geopolymer. Results showed that the pore structure of

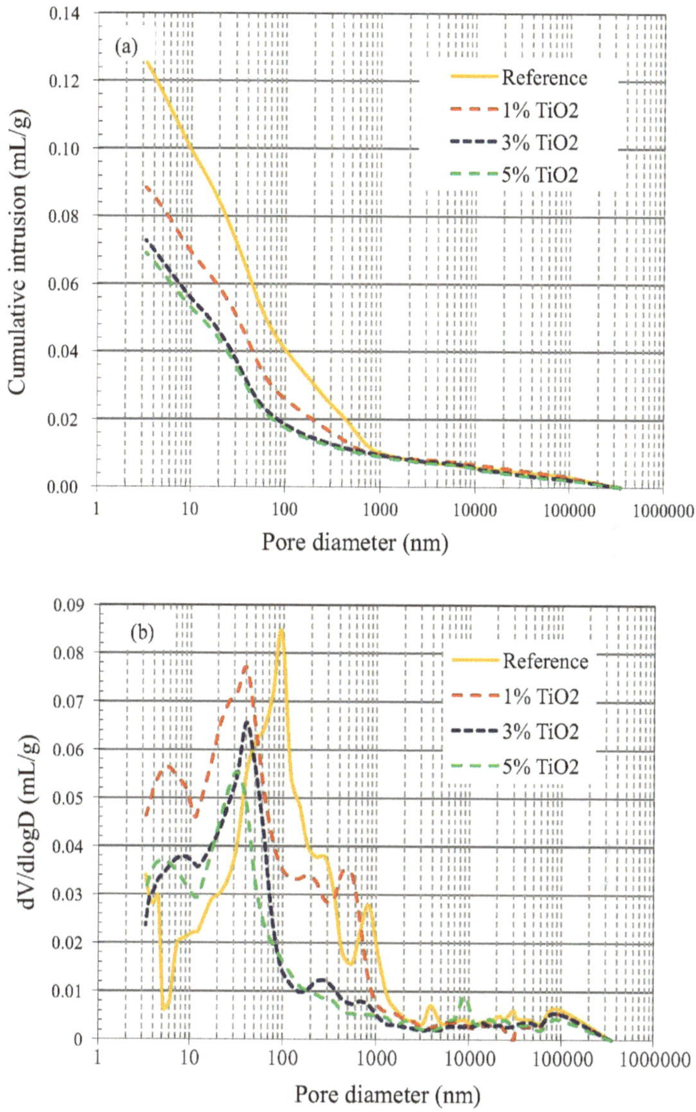

Fig. 7.5 Pore size distribution and porosity of FA-based geopolymer with nanotitania [49].

mixtures with nanotitania was significantly smaller compared to control samples. With the addition of different sizes of nanotitania particles, specific pore volumes of the geopolymers were reduced. The most probable pore diameter of the geopolymer was decreased with the decreased size of nanotitania. The pore diameter of the samples shifted to the range of the less-harmful pores [63] by the filler effect of nanotitania, which further increased the geopolymer density.

For activated slag, a study by Yang et al. [57] reported differences in microstructure between the reference alkali-activated slag and those with TiO_2 nanoparticles. The SEM images showed that the incorporation of nano-TiO_2 accelerated the hydration process, resulting in more hydration products like C—S—H/C—S—A—H and a densified microstructure. The results of cumulative porosity also showed that nano-TiO_2 lowered the total porosity. This revealed that nano-TiO_2 changed the pore structure toward a more mesopore structure (1.25–25 nm).

7.4 Effect of nanomaterials on fresh and hardened properties of geopolymers

7.4.1 Workability and setting time

7.4.1.1 Nanosilica

Revathi et al. [64] studied the effect of NS on the early-age properties of slag/FA-based geopolymer. The results showed that NS reduced the flowability of the mixtures due to the high surface area and very fine particle size of NS. Gülsan et al. [65] investigated the simultaneous effect of NS and steel fiber (SF) on the fresh properties of self-consolidating geopolymer using 50% FA and 50% ground granulated blast furnace slag. The test results at fresh state showed that the use of NS and SF had a negative effect on the slump flow diameter, t_{50} flow time, V-funnel flow time, and L-Box passing ability. The incorporation of 2% NS reduced the fresh property values in the geopolymer. However, NS improved the resistance to segregation and bleeding by increasing the cohesiveness of the mixtures. Shahrajabian and Behfarnia [66] reported that NS reduced the slump of alkali-activated slag. This was related to the smaller size of nanoparticles compared to slag with a larger aspect ratio that resulted in higher water absorption and slump reduction. The fineness of NS and the increased water demand was also reported in [67]. Berra et al. [68] recommended using a superplasticizer (polyacrylic type) to overcome the slump reduction. However, the immediate addition of superplasticizer after mixing water containing NS had adverse effects on the effectiveness of NS. The authors related such adverse effects to the interaction between superplasticizer and NS that lowered the reactivity of nanoparticles. They recommended that a delayed addition of the superplasticizer will help the uniform dispersion of materials without significantly affecting the NS reactivity. Xie and Kayali [69] reported that polycarboxylate superplasticizers lowered mixture stiffness and improved the flowability of FA-based geopolymer.

Some studies showed that the accelerating effect of NS on geopolymers can lead to increased workability. Adak et al. [35] reported that increasing the NS content increased the slump of FA-based geopolymer. Rodríguez et al. [70] explained this phenomenon by indicating that the incomplete dissolution of NS in the hydroxide solution decreases the viscosity and improves workability. Deb et al. [33] also reported acceptable workability and easy handling of mixtures with NS during the casting of the samples.

Adak et al. [35] studied an FA-based geopolymer that was activated with NaOH with three molar concentrations of 8 (M), 10 (M), and 12 (M) and was mixed with Na_2SiO_3 solution in the proportion of 1:1.75 (by mass). In their study, colloidal NS with up to 10% of FA, by mass, was applied. The results showed that the addition of NS reduced the initial and setting time of the FA geopolymer from 200 and 274 min to 113 and 162 min when 4% NS was used. A similar trend was observed in MK-based geopolymer by Gao et al. [36]. The authors showed that adding 1% NS (with an average size of 10 nm) reduced the initial and final setting times. They reported that increased sodium silicate content increased the paste's viscosity and shortened the setting time. Zidi et al. [71] also noted that adding 5% NS reduced the setting time of MK-based geopolymer from 360 min by 45%. The authors attributed the such reduction in setting time to high specific surface area and high reactivity of NS. Therefore, it can be stated that the inclusion of NS to low calcium-bearing procurers can reduce the setting of geopolymers and facilitate their hardening at ambient room temperature. This can help overcome the issue caused by heat curing conditions. However, higher contents of NS caused agglomeration and reduced the accelerating effect of NS and the associated influence on the setting time. For FA with high-calcium content, Phoo-ngernkham et al. [51] reported that NS to Class C FA geopolymer resulted in decreased setting time due to the increased formation of C—S—H.

7.4.1.2 Nanoclay

Shahrajabian and Behfarnia [66] investigated the effect of nanoclay particles on the workability of alkali-activated slag. They reported the workability was not affected significantly comparing to those with NS and nanoalumina. They mentioned that nanoclay remained nonreactive in contact with water and the alkaline solution, followed by an insignificant change in the slump. Hassaan et al. [72] reported that the slag/MK-geopolymer with 3% nanoclay obtained good workability. The mixture was further enhanced when a 4% Glenium ACE-based superplasticizer was added.

7.4.1.3 Nanoalumina

Phoo-ngernkham et al. [51] reported that the setting time of FA-based geopolymer could be slightly affected (reduced) by nanoalumina. In their study, high-calcium FA was activated using 10M sodium hydroxide and sodium silicate at a constant liquid-to-binder ratio of 0.60 and $Na_2SiO_3/NaOH$ ratio of 2.0. The nanoalumina with particle sizes of 13nm was added at the dosages of 0%, 1%, 2%, and 3% by weight. They reported that the effect of nanoalumina was less than that of NS in geopolymer mixtures with a similar design (and identical dosage). It was due to the higher reactivity of NS particles, especially with calcium ions found in abundance in high-calcium FA. Fig. 7.6 compares the effect of NS (shown as S) and nanoalumina (shown as A) on the initial and final setting times of FA-based geopolymer. As can be seen, the addition of 3% NS decreased the setting time by 50%, while nanoalumina had a very slight effect on the setting times. However, Chindaprasirt et al. [61] reported that the initial and final setting times of high-calcium FA-based geopolymer were reduced with the addition of 7% nanoalumina. Still, an increase in nanoalumina accelerated the setting of high-calcium FA geopolymer.

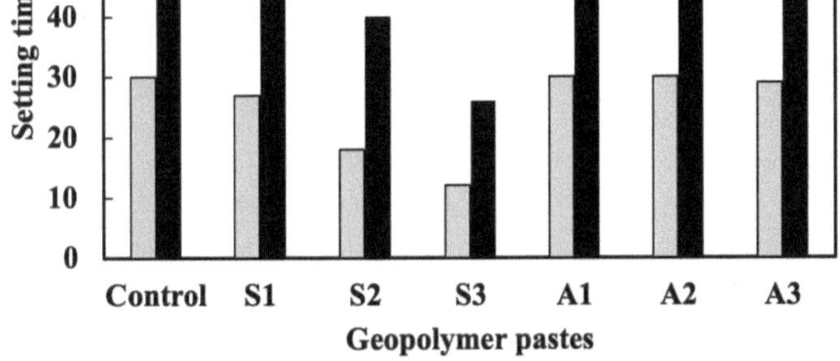

Fig. 7.6 Setting time of high-calcium FA-based geopolymer with NS and nanoalumina [51].

7.4.1.4 Nanotitania

Like nanoalumina, nanotitania performance in geopolymer is mainly based on the nano-sized particles that facilitate nucleation and filling effects. Physical characteristics of nanotitania can shorten the setting time and lower the workability of geopolymers. Guzmán-Aponte et al. [55] showed that the setting time of MK-based geopolymer was reduced with the inclusion of TiO_2. Also, Duan et al. [49] reported that 5% nanotitania decreased the workability of FA-based geopolymer by 22%.

7.4.2 Mechanical properties

7.4.2.1 Nanosilica

NS can contribute to the densification of the geopolymer through physical and chemical effects. This property is unique among nanoparticles as other particles such as nanoalumina and nanotitania mainly possess physical effects owing to their crystalline nature. The literature reports a significant improvement in compressive strength and flexural strength of geopolymers with NS regardless of precursors. Phoo-ngernkham et al. [51] reported higher compressive and flexural strengths of 52 MPa and 6 MPa at 90 days, respectively, for high-calcium FA-based geopolymer with the addition of 2% NS. It was equal to an increase of 32% and 38% as compared with the control samples, respectively. They attributed such enhancement to the formation of secondary geopolymeric gel (C—S—H/C—A—S—H and N—A—S—H). Results showed that NS reacted with the calcium ions from high-calcium FA, which led to the formation of secondary strengthening gel and enhanced the strength. Another study by Deb et al. [58] on FA-based geopolymer showed the highest increase in compressive strength with 3% NS at all ages up to 90 d. Both studies referred to the agglomeration of the excessive NS that resulted in impaired mechanical properties.

Wang et al. [39] reported the effect of NS (dosage of 0.5%–3%) on the alkali-activated slag. They noted that an increase in NS (up to 2%) increased the compressive strength. Any content higher than 2% reduced the compressive strength. It was also reported that the early-age compressive strength was better than the later age strength due to the high reactivity of NS and slag. The threshold for inclusion of NS was reported to be 1% for MK-based geopolymer. Gao et al. [36] mentioned that the addition of more than 1% NS reduced the strength of geopolymer. The authors reported that the excessive amount of NS leached out of the matrix. The variation in the threshold of NS can be related to the content of Ca ions of aluminosilicates. The Ca ions in the aluminosilicates can react with NS and produce more strengthening gel. In a study by Zidi et al. [71], it was shown that up to 5% NS contributed to the higher compressive strength of MK-based geopolymer cured at 80°C as compared to the control mixtures. However, the higher content of NS with better performance in their study can be related to the increase in the curing temperature. Temperature can increase the reactivity of the mixture.

Higher content of NS in colloidal form was also reported to be influential on the mechanical properties of geopolymers. Adak et al. [35] investigated the effect of 6% colloidal NS on the mechanical properties of FA-based geopolymer, where a significant increase was reported. The results showed that the added colloidal NS increased the Si/Al and added to the homogeneity of the matrix. Nanomodification of the matrix developed a denser interfacial transition zone (ITZ) between aggregates and the matrix, which enhanced bond strength and flexural strengths of geopolymer. The enhancing performance of NS on the ITZ was also reported by Gulsan et al. [65]. The authors studied the hybrid effect of NS and SF on the mechanical properties of self-consolidating geopolymer concretes (SCGC). It was reported that the incorporation of NS in the SCGC mixtures did not significantly affect flexural performance when used without fibers. However, the hybrid uses of NS (2%) and SFs (1%) enhanced bond strength, flexural strength, and fracture properties of the geopolymer due to NS's positive effect on the ITZ of fibers. Fig. 7.7 shows the load-deflection responses of slag/FA-based geopolymer

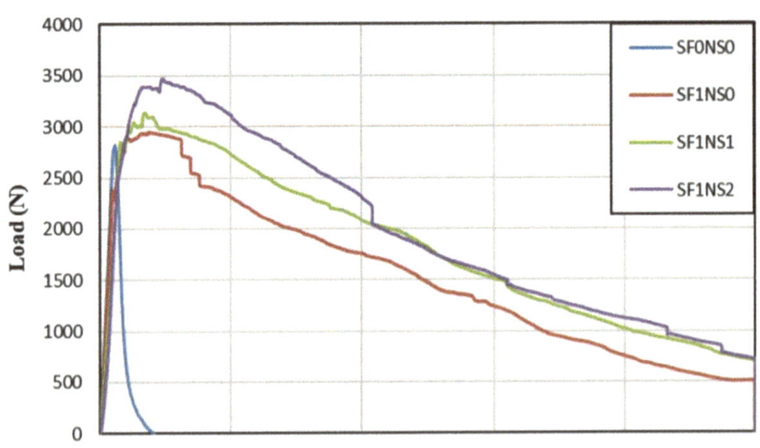

Fig. 7.7 Load deflection responses of 1% SF and different contents of NS in slay/F-based geopolymer [65].

with 0.5% SF and NS (1% and 2%). As seen, the samples with NS showed more strain hardening behavior due to higher bond strength. Xu et al. [73] also confirmed such performance in FA-based geopolymers with up to 6% NS and polyvinyl alcohol (PVA) fibers. The authors reported that the mechanical properties of the geopolymer mixtures were increased with NS cured at ambient temperature. The optimum content of 6% NS and 5% PVA fibers was recommended to achieve the optimum mechanical properties of FA-based geopolymer.

7.4.2.2 Nanoclay

Similar to NS, nanoclay can significantly improve the mechanical properties of geopolymers. However, the optimum content of nanoclay can vary in different aluminosilicate precursors based on their chemical composition. Assaaedi et al. [42] reported that 2% nanoclay is the optimum content to obtain the maximum flexural and compressive strengths, flexural modulus, and hardness of FA-based geopolymer. This threshold was also limited to 2% for alkali-activated slag, and higher contents reduced the compressive strength at 7 and 28 days [66]. However, In another study by Kaur et al. [74], 4% replacement of nanoclay was reported as the optimum content for FA-based geopolymer. The SEM examinations showed that silica and alumina oxide of nanoclay contributed to the higher polymerization reaction. SEM analysis showed that the geopolymer with nanoclay (lower contents) had a less dense microstructure. It was related to the presence of a higher number of unreacted and partially reacted FA particles. Also, results from EDS analysis showed that the optimum content of nanoclay reduced the Si/Al of geopolymer, showing more involvement of Si in the matrix and a higher degree of polymerization that increased the strength. In a study by Revitheja and Kumar [75], it was reported that the addition of 6% nanoclay in FA-based geopolymer, cured at ambient temperature, improved the compressive strength, splitting tensile strength, and flexural strength by 26%, 29%, and 37%, respectively. The authors reported that such enhancement was mainly attributed to the high surface area of nanoclay that enhanced the degree of polymerization. The excessive addition of nanoclay (above 6%) did not lead to the active participation of nanoparticles in the reaction process, and the beneficial effect was strength was insignificant. For the slag/FA-based geopolymer, the improvement was increased to 47%, 51%, and 48% for compressive strength, splitting tensile strength, and flexural strength, respectively, compared to mixtures without nanoclay.

7.4.2.3 Nanoalumina

Nanoalumina also has a beneficial effect on the mechanical properties of geopolymer. The threshold for inclusion of nanoalumina was reported to be 2%. However, the enhancing mechanism of alumina nanoparticles is different from that of NS. The positive impact of nanoalumina is mainly originated from the crystallinity of alumna nanoparticles. Similar to other nanoparticles, an excessive amount of inclusion would cause agglomeration and reduce the strength of geopolymers. Alomayri [48] studied the effect of 1% to 3% nanoalumina particles by weight into FA-based geopolymer with a constant liquid-to-binder ratio of 0.75 cured at ambient room temperature. Results showed that the optimum content of nanoalumina particles was 2% for improving the mechanical properties of geopolymer, where a maximum improvement of approximately 45% for flexural strength was recorded. The enhanced flexural strength was attributed to the improvement in the matrix by the filling effect of nanoparticles, creating a denser microstructure [40]. This effect led to overcoming local failures in the sample by enhancing the interfacial adhesion, leading to superior resistance to bending and fracture forces. The authors reported that nanoalumina particles bridged the microcracks and inhibited crack propagation.

Phoo-ngernkham et al. [51] studied the effect of nanoalumina on high-calcium FA-based geopolymer at contents of 1%, 2%, and 3%. They found that nanoalumina particles enhanced the compressive and flexural strengths at ambient temperature compared to the reference mixtures. The compressive strength, flexural strength, and elastic modulus of high-calcium FA-based geopolymer with nanoalumina increased due to the higher formation rate of C—S—H/C—A—S—H and N—A—S—H. The formation of geopolymer gels can fill the pore to make the dense and robust geopolymer. However, higher content of nanoalumina was needed (2%) to obtain similar results for mixtures with NS (1%). In another study by Shahrajabian and Behfarnia [66], the effect of nanoalumina on the compressive strength of alkali-activated slag was investigated. They reported that the increase of nanoalumina by 2% increased the compressive strength. However, the compressive strength of mixtures with 3% nanoalumina decreased. The authors related this decrease to the agglomeration and uneven distribution of nanoparticles during the mixing process. However, it was suggested that nanoalumina is not a promising solution for improving the compressive strength of alkali-activated slag due to the high cost of these particles.

7.4.2.4 Nanotitania

Duan et al. [49] studied the effect of 1%, 3%, and 5% nanotitania on the compressive strength of fluidized bed FA-based geopolymer concrete. The authors reported the enhancing effect of nanotitania on compressive strength at both early and later ages. As reported, the enhancing effect of nanotitania was more accentuated due to a faster polymerization rate at an early age. Maiti et al. [62] reported that 5% nanotitania was the optimum amount for maximum compressive strength (53 MPa) and split-tensile strength (6.8 MPa) after 28 days of curing. Nanotitania particles reduced the porosity, roughness and increased the density of the geopolymer matrix. For alkali-activated slag, though, Yang et al. [57] limited the incorporation of nanotitania to 0.5% and reported that the compressive strength improved by 10%, 15%, and 9% compared to the reference mixtures at 3, 7, and 28 days, respectively. In contrast, flexural strength improved 25%, 25%, and 38% for corresponding 3, 7, and 28 days. Results from Llano-Guerrero et al. [56] also indicated that slag/Mk-based geopolymer with 0.5% of TiO_2 had higher compressive strength. Nanoparticles acted as fillers in a denser reaction products matrix, helping to reduce cracks and therefore improving the compressive strength.

7.4.3 Durability

7.4.3.1 Nanosilica

One of the significant roles of NS is improving the impermeability of geopolymers. Deb et al. [58] studied the sorptivity of geopolymers with 0% and 2% NS. Geopolymer mortars using FA, binary mixtures of FA and 10% Portland cement, and FA and 15% slag were prepared. Results were indicative of the reduction in the sorptivity coefficient for all mixtures when NS was applied. They attributed such enhancement to the densification of the microstructure through the nanofilling effect of particles as well as the formation of additional geopolymer gel. However, NS showed the lowest performance in the FA-based geopolymer. Zidi et al. [71] showed that the addition of 5% NS into the geopolymer mixture increased the density from 1.62 to 1.72 g/cm^{-3} and contributed to the decrease in the water absorption of the geopolymer by about 10.5%. The authors attributed the improvement to the formation of the homogeneous structure of the geopolymer. The SEM images also confirmed an accumulation of the gel and the filling of the pores. Gao et al. [76] reported that adding 1% to 3% NS to slag/FA-based geopolymer reduced the water permeability of the mixtures by 26%.

However, the addition of any excessive content of NS increased the permeability by generating air voids, mainly caused by the adverse effect of nanoparticles on the workability of the mixture. Their et al. [77] investigated the effect of NS on gas permeability of slag/FA-based geopolymer incorporated fiber and coal fly ash aggregate (CFA). The incorporation of CFA increased the permeability of the system due to its intrinsic porous structure. The authors reported that NS improved the water penetration resistance, gas permeability, and sorptivity of CFA mixtures. The refining of pore structure and discontinuity of pores resulting from NS decreased the gas permeability coefficient. Also, Saini and Vattipalli [78] studied the durability of self-consolidating alkali-activated slag with 2% NS through sorptivity index value. They indicated that 2% NS provided the highest strength and durability performance without compromising its workability.

Ramezanianpour and Moeini [79] studied the rapid chloride migration in alkali-activated slag with 2% to 4% NS and up to 10% silica fume. It was reported that NS addition contributed to further gel production and lower porosity, inhibiting the chloride ingression into the matrix. Cevik et al. [80] investigated the effect of NS on the short-term severe durability performance of FA-based geopolymer concrete. The results showed that 3% NS enhanced the resistance of the mixtures to chemical environments (sulfuric acid solution and seawater). Fig. 7.8 shows the residual flexural strength of FA-based geopolymer with NS after exposure to the chemical environment. As seen, the strength reduction after exposure was minimum for mixtures with NS. The results also showed denser structure and decreased porosity and permeability of mixtures with NS.

7.4.3.2 Nanoclay

Assaedi et al. [42] studied the effect of using 1% to 3% nanoclay on the porosity and water absorption of the FA-based geopolymer. Results showed that the addition of nanoclay reduced the porosity and water absorption. Incorporating 2% nanoclay was reported to bring about the optimum performance by decreasing the porosity (7%) and water absorption (17%) and 17%. Shahrajabian and Behfarnia [66] investigated the effect of nanoclay on the frost resistance of alkali-activated slag. Results were indicative that 2% nanoclay enhanced the residual compressive strength of mixtures after exposure to negative freeze and thaw cycles. Also, the exposed nanomodified samples experienced less mass loss after exposure. The results were in good agreement with another study [51].

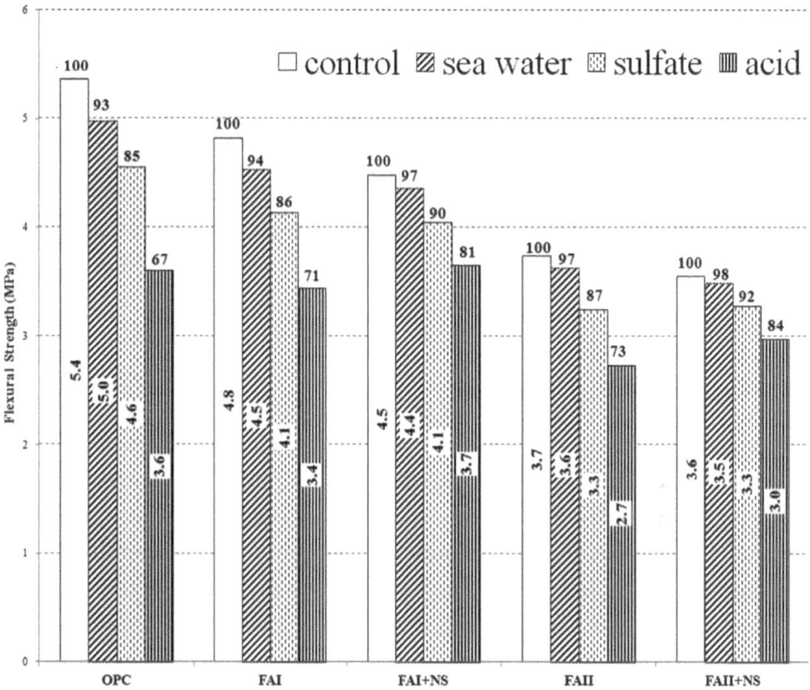

Fig. 7.8 Residual flexural strength of FA-based geopolymer with NS after exposure to the chemical environment [80].

7.4.3.3 Nanoalumina

Very few studies have investigated the effect of nanoalumina on the durability of geopolymers. Shahrajabian and Behfarnia [66] studied the durability of alkali-activated slag with the presence of nanoalumina against freeze and thaw cycles. The results showed that the addition of 2% nanoalumina reduced compressive strength loss of geopolymer under freezing-thawing cycles. However, the effect of nanoalumina was not significant in preventing the mass loss of mixtures under freeze and thaw cycles. They reported that incorporating NS and nanoclay was more effective than nanoalumina on the frost durability resistance of geopolymers.

7.4.3.4 Nanotitania

Maiti et al. [62] investigated the effect of nanotitania on the apparent volume of permeable voids and water absorption of FA-based geopolymer. Results showed that 5% nanotitania reduced the volume of available permeable voids in the matrix. Yang et al. [57] also reported that the addition of

0.5% nanotitania in alkali-activated slag reduced the total porosity of pore size (1.25–25 nm) and improved the pore size distribution. Both studies found that nanotitania filled the nanopores in the geopolymer matrix and increased resistance against deleterious agents. Duan et al. [49] reported that the nanotitania is also effective in enhancing the carbonation resistance of geopolymers. Their results showed that carbonation depth decreased with the incorporation of nanotitania under accelerated carbonation. They reported that 5% nanotitania was the optimum content and reduced the carbonation depth by 100%.

7.5 Summary and perspective

This chapter discussed the effect of the most widely used nanomaterials, namely, NS, nanoclay, nanoalumina, and nanotitania, on reaction kinetics, geopolymer products, microstructure, and properties of geopolymers. Similar to their functions in Portland cement systems, nanoparticles can accelerate the reaction process due to their high surface energy, high reactivity, and chemical instability, improving their active participation in polymerization. This can increase the dissolution of aluminosilicates, nucleation and growth of initial solid phases, and interactions of formed phases at the boundaries.

The addition of nanomaterials increased the dissolution rate of silica and alumina phases in the precursor and promoted the formation of long-chain silicate oligomers. Also, nanomaterials can act as nucleation sites in the matrix leading to an even distribution of geopolymer products in the matrix. The chemical characterization of geopolymers indicates that the size and amount of unreacted particles decrease with the addition of nanomaterials. However, studies have shown that agglomeration of nanoparticles can happen when higher contents are applied, which causes the formation of voids. Therefore, the homogeneity of the microstructure will be lowered.

Nanomaterials play a crucial role in both the fresh and hardened properties of geopolymers. Studies show that the incorporation of nanomaterials can decrease the setting time of geopolymers. This is due to the increased formation of geopolymer products. Such accelerated setting can lower the workability of geopolymers, which necessitates the use of superplasticizers at appropriate contents and order of addition. Despite the adverse effect of some nanomaterials on workability, they increase the mechanical properties of geopolymers. They enhance geopolymers' bond strength and flexural strength due to developing a stronger ITZ between aggregates/fibers and the bulk geopolymer paste. The nanomaterials also improve the impermeability of microstructure and densify the matrix. This can increase the resistance of geopolymers to chloride ingression, carbonation, and other deterioration mechanisms.

References

[1] C. Shi, D. Roy, P. Krivenko, Alkali-Activated Cements and Concretes, CRC Press, 2003.

[2] E. Álvarez-Ayuso, X. Querol, F. Plana, A. Alastuey, N. Moreno, M. Izquierdo, O. Font, T. Moreno, S. Diez, E. Vázquez, Environmental, physical and structural characterization of geopolymer matrixes synthesised from coal (co-) combustion fly ashes, J. Hazard. Mater. 154 (2008) 175–183.

[3] H. Kuhl, Slag Cement and Process of Making the Same, 1908.

[4] V. Glukhovsky, Soil Silicates, vol. 154, Gosstroyizdat, Kiev, 1959.

[5] J. Davidovits, Geopolymers: inorganic polymeric new materials, J. Therm. Anal. Calorim. 37 (1991) 1633–1656.

[6] J.L. Provis, S.A. Bernal, Geopolymers and related alkali-activated materials, Annu. Rev. Mat. Res. 44 (2014) 299–327.

[7] J.L. Provis, Geopolymers and other alkali activated materials: why, how, and what? Mater. Struct. 47 (2014) 11–25.

[8] J.L. Provis, A. Palomo, C. Shi, Advances in understanding alkali-activated materials, Cem. Concr. Res. 78 (2015) 110–125.

[9] C. Shi, A.F. Jiménez, A. Palomo, New cements for the 21st century: the pursuit of an alternative to Portland cement, Cem. Concr. Res. 41 (2011) 750–763.

[10] S. Puligilla, P. Mondal, Role of slag in microstructural development and hardening of fly ash-slag geopolymer, Cem. Concr. Res. 43 (2013) 70–80.

[11] J.L. Provis, J.S. Van Deventer, Alkali Activated Materials: State-of-the-Art Report, RILEM TC 224-AAM, Springer Science & Business Media, 2013.

[12] M. Criado, A. Palomo, A. Fernández-Jiménez, Alkali activation of fly ashes. Part 1: effect of curing conditions on the carbonation of the reaction products, Fuel 84 (2005) 2048–2054.

[13] A. Kusbiantoro, M.S. Ibrahim, K. Muthusamy, A. Alias, Development of sucrose and citric acid as the natural based admixture for fly ash based geopolymer, Procedia Environ. Sci. 17 (2013) 596–602.

[14] J.G. Jang, N. Lee, H.-K. Lee, Fresh and hardened properties of alkali-activated fly ash/slag pastes with superplasticizers, Construct. Build Mater. 50 (2014) 169–176.

[15] Y. Ling, K. Wang, C. Fu, Shrinkage behavior of fly ash based geopolymer pastes with and without shrinkage reducing admixture, Cem. Concr. Compos. 98 (2019) 74–82.

[16] K.T. Nguyen, T.A. Le, J. Lee, D. Lee, K. Lee, Investigation on properties of geopolymer mortar using preheated materials and thermogenetic admixtures, Construct. Build Mater. 130 (2017) 146–155.

[17] N. Ali, O. Canpolat, Y. Aygörmez, M.M. Al-Mashhadani, Evaluation of the 12–24 mm basalt fibers and boron waste on reinforced metakaolin-based geopolymer, Construct. Build Mater. 251 (2020), 118976.

[18] G.F. Huseien, J. Mirza, M. Ismail, S. Ghoshal, M.A.M. Ariffin, Effect of metakaolin replaced granulated blast furnace slag on fresh and early strength properties of geopolymer mortar, Ain Shams Eng. J. 9 (2018) 1557–1566.

[19] B. Panda, M.J. Tan, Experimental study on mix proportion and fresh properties of fly ash based geopolymer for 3D concrete printing, Ceram. Int. 44 (2018) 10258–10265.

[20] F. Nuruddin, S. Demie, F.A. Memon, N. Shafiq, Effect of superplasticizer and NaOH molarity on workability, compressive strength and microstructure properties of self-compacting geopolymer concrete, World Acad. Sci. Eng. Technol. 75 (2011).
[21] C. Montes, E. Allouche, Rheological behavior of fly-ash-based geopolymers, in: Geopolymer Binder Systems, ASTM International, 2013.
[22] P. Zhang, K. Wang, Q. Li, J. Wang, Y. Ling, Fabrication and engineering properties of concretes based on geopolymers/alkali-activated binders-a review, J. Clean. Prod. (2020), 120896.
[23] M. Sofi, J. Van Deventer, P. Mendis, G. Lukey, Engineering properties of inorganic polymer concretes (IPCs), Cem. Concr. Res. 37 (2007) 251–257.
[24] B. Singh, G. Ishwarya, M. Gupta, S. Bhattacharyya, Geopolymer concrete: a review of some recent developments, Construct. Build Mater. 85 (2015) 78–90.
[25] M. Albitar, M.M. Ali, P. Visintin, M. Drechsler, Durability evaluation of geopolymer and conventional concretes, Construct. Build Mater. 136 (2017) 374–385.
[26] A. Fernández-Jiménez, F. Puertas, A. Arteaga, Determination of kinetic equations of alkaline activation of blast furnace slag by means of calorimetric data, J. Therm. Anal. Calorim. 52 (1998) 945–955.
[27] S.A. Bernal, J.L. Provis, V. Rose, R.M. De Gutierrez, Evolution of binder structure in sodium silicate-activated slag-metakaolin blends, Cem. Concr. Compos. 33 (2011) 46–54.
[28] J.I. Escalante-García, A.F. Fuentes, A. Gorokhovsky, P.E. Fraire-Luna, G. Mendoza-Suarez, Hydration products and reactivity of blast-furnace slag activated by various alkalis, J. Am. Ceram. Soc. 86 (2003) 2148–2153.
[29] A. Fernández-Jiménez, F. Puertas, I. Sobrados, J. Sanz, Structure of calcium silicate hydrates formed in alkaline-activated slag: influence of the type of alkaline activator, J. Am. Ceram. Soc. 86 (2003) 1389–1394.
[30] J. Provis, S. Yong, P. Duxson, Nanostructure/microstructure of metakaolin geopolymers, in: Geopolymers, Elsevier, 2009, pp. 72–88.
[31] A. Fernández-Jiménez, A. Palomo, Nanostructure/microstructure of fly ash geopolymers, in: Geopolymers, Elsevier, 2009, pp. 89–117.
[32] M.-H. Zhang, J. Islam, S. Peethamparan, Use of nano-silica to increase early strength and reduce setting time of concretes with high volumes of slag, Cem. Concr. Compos. 34 (2012) 650–662.
[33] P.S. Deb, P.K. Sarker, S. Barbhuiya, Effects of nano-silica on the strength development of geopolymer cured at room temperature, Construct. Build Mater. 101 (2015) 675–683.
[34] M. Ibrahim, M.A.M. Johari, M. Maslehuddin, M.K. Rahman, Influence of nano-SiO_2- on the strength and microstructure of natural pozzolan based alkali activated concrete, Construct. Build Mater. 173 (2018) 573–585.
[35] D. Adak, M. Sarkar, S. Mandal, Structural performance of nano-silica modified fly-ash based geopolymer concrete, Construct. Build Mater. 135 (2017) 430–439.
[36] K. Gao, K.-L. Lin, D. Wang, C.-L. Hwang, H.-S. Shiu, Y.-M. Chang, T.-W. Cheng, Effects SiO_2/Na_2O molar ratio on mechanical properties and the microstructure of nano-SiO_2 metakaolin-based geopolymers, Construct. Build Mater. 53 (2014) 503–510.
[37] Y.-L. Tsai, J.V. Hanna, Y.-L. Lee, M.E. Smith, J.C. Chan, Solid-state NMR study of geopolymer prepared by sol–gel chemistry, J. Solid State Chem. 183 (2010) 3017–3022.
[38] K.-W. Lo, K.-L. Lin, T.-W. Cheng, Y.-M. Chang, J.-Y. Lan, Effect of nano-SiO2 on the alkali-activated characteristics of spent catalyst metakaolin-based geopolymers, Construct. Build Mater. 143 (2017) 455–463.
[39] J. Wang, P. Du, Z. Zhou, D. Xu, N. Xie, X. Cheng, Effect of nano-silica on hydration, microstructure of alkali-activated slag, Construct. Build Mater. 220 (2019) 110–118.

[40] A. Hakamy, F. Shaikh, I.M. Low, Characteristics of nanoclay and calcined nanoclay-cement nanocomposites, Compos. Part B 78 (2015) 174–184.
[41] M. Aly, M. Hashmi, A. Olabi, M. Messeiry, A. Hussain, Effect of nano clay particles on mechanical, thermal and physical behaviours of waste-glass cement mortars, Mater. Sci. Eng. A 528 (2011) 7991–7998.
[42] H. Assaedi, F. Shaikh, I.M. Low, Effect of nano-clay on mechanical and thermal properties of geopolymer, J. Asian Ceramic Soc. 4 (2016) 19–28.
[43] A. Nazari, J.G. Sanjayan, Hybrid effects of alumina and silica nanoparticles on water absorption of geopolymers: application of Taguchi approach, Measurement 60 (2015) 240–246.
[44] Y. Huang, M. Han, The influence of α-Al_2O_3 addition on microstructure, mechanical and formaldehyde adsorption properties of fly ash-based geopolymer products, J. Hazard. Mater. 193 (2011) 90–94.
[45] S. Riahi, A. Nazari, The effects of nanoparticles on early age compressive strength of ash-based geopolymers, Ceram. Int. 38 (2012) 4467–4476.
[46] A. Hajimohammadi, J.L. Provis, J.S. van Deventer, The Effect of Nanoparticle Seeding on the Mechanism of One-Part Geopolymer Gel Formation, 2009.
[47] C.A. Rees, J.L. Provis, G.C. Lukey, J.S. Van Deventer, The mechanism of geopolymer gel formation investigated through seeded nucleation, Colloids Surf. A Physicochem. Eng. Asp. 318 (2008) 97–105.
[48] T. Alomayri, Experimental study of the microstructural and mechanical properties of geopolymer paste with nano material (Al2O3), J. Build. Eng. 25 (2019), 100788.
[49] P. Duan, C. Yan, W. Luo, W. Zhou, Effects of adding nano-TiO_2 on compressive strength, drying shrinkage, carbonation and microstructure of fluidized bed fly ash based geopolymer paste, Construct. Build Mater. 106 (2016) 115–125.
[50] K. Gao, K.-L. Lin, D. Wang, C.-L. Hwang, B.L.A. Tuan, H.-S. Shiu, T.-W. Cheng, Effect of nano-SiO_2 on the alkali-activated characteristics of metakaolin-based geopolymers, Construct. Build Mater. 48 (2013) 441–447.
[51] T. Phoo-ngernkham, P. Chindaprasirt, V. Sata, S. Hanjitsuwan, S. Hatanaka, The effect of adding nano-SiO_2 and nano-Al_2O_3 on properties of high calcium fly ash geopolymer cured at ambient temperature, Mater. Des. 55 (2014) 58–65.
[52] K. Behfarnia, M. Rostami, Effects of micro and nanoparticles of SiO_2 on the permeability of alkali activated slag concrete, Construct. Build Mater. 131 (2017) 205–213.
[53] J.D. Grice, The structure of spurrite, Tilleyite and scawtite, and relationships to other silicate–carbonate minerals, Can. Mineral. 43 (2005) 1489–1500.
[54] H. Assaedi, F. Shaikh, I.M. Low, Effect of nanoclay on durability and mechanical properties of flax fabric reinforced geopolymer composites, J. Asian Ceramic Soc. 5 (2017) 62–70.
[55] L.A. Guzmán-Aponte, R. Mejia de Gutierrez, A. Maury-Ramírez, Metakaolin-based geopolymer with added TiO_2 particles: physicomechanical characteristics, Coatings 7 (2017) 233.
[56] E. Llano-Guerrero, L. Gómez-Zamorano, E. Jiménez-Relinque, Effect of the addition of TiO_2 nanoparticles in alkali-activated materials, Construct. Build Mater. 245 (2020), 118370.
[57] L. Yang, Z. Jia, Y. Zhang, J. Dai, Effects of nano-TiO_2 on strength, shrinkage and microstructure of alkali activated slag pastes, Cem. Concr. Compos. 57 (2015) 1–7.
[58] P.S. Deb, P.K. Sarker, S. Barbhuiya, Sorptivity and acid resistance of ambient-cured geopolymer mortars containing nano-silica, Cem. Concr. Compos. 72 (2016) 235–245.
[59] H. Assaedi, F. Shaikh, I.M. Low, Influence of mixing methods of nano silica on the microstructural and mechanical properties of flax fabric reinforced geopolymer composites, Construct. Build Mater. 123 (2016) 541–552.

[60] H. Khater, B. El-Sabbagh, M. Fanny, M. Ezzat, M. Lottfy, Effect of nano-silica on alkali activated water cooled slag geopolymer, ARPN J. Eng. Appl. Sci. 2 (2012) 170–176.
[61] P. Chindaprasirt, P. De Silva, K. Sagoe-Crentsil, S. Hanjitsuwan, Effect of SiO_2 and Al_2O_3 on the setting and hardening of high calcium fly ash-based geopolymer systems, J. Mater. Sci. 47 (2012) 4876–4883.
[62] M. Maiti, M. Sarkar, S. Maiti, M.A. Malik, S. Xu, Modification of geopolymer with size controlled TiO_2 nanoparticle for enhanced durability and catalytic dye degradation under UV light, J. Clean. Prod. 255 (2020), 120183.
[63] M. Zhang, H. Li, Pore structure and chloride permeability of concrete containing nano-particles for pavement, Construct. Build Mater. 25 (2011) 608–616.
[64] T. Revathi, R. Jeyalakshmi, N. Rajamane, Study on the role of n-SiO_2 incorporation in thermo-mechanical and microstructural properties of ambient cured FA-GGBS geopolymer matrix, Appl. Surf. Sci. 449 (2018) 322–331.
[65] M.E. Gülşan, R. Alzeebaree, A.A. Rasheed, A. Niş, A.E. Kurtoğlu, Development of fly ash/slag based self-compacting geopolymer concrete using nano-silica and steel fiber, Construct. Build Mater. 211 (2019) 271–283.
[66] F. Shahrajabian, K. Behfarnia, The effects of nano particles on freeze and thaw resistance of alkali-activated slag concrete, Construct. Build Mater. 176 (2018) 172–178.
[67] G.Q. Bianchi, H. Brouwers, Application of Nano-Silica (nS) in Concrete Mixtures, Technical University of Denmark, 2010, pp. 431–436.
[68] M. Berra, F. Carassiti, T. Mangialardi, A. Paolini, M. Sebastiani, Effects of nanosilica addition on workability and compressive strength of Portland cement pastes, Construct. Build Mater. 35 (2012) 666–675.
[69] J. Xie, O. Kayali, Effect of superplasticiser on workability enhancement of class F and class C fly ash-based geopolymers, Construct. Build Mater. 122 (2016) 36–42.
[70] E.D. Rodríguez, S.A. Bernal, J.L. Provis, J. Paya, J.M. Monzo, M.V. Borrachero, Effect of nanosilica-based activators on the performance of an alkali-activated fly ash binder, Cem. Concr. Compos. 35 (2013) 1–11.
[71] Z. Zidi, M. Ltifi, I. Zafar, Synthesis and attributes of nano-SiO_2 local metakaolin based-geopolymer, J. Build. Eng. 33 (2021) 101586.
[72] M.M. Hassaan, H.M. Khater, M.S. El-Mahllawy, A.M. El Nagar, Production of geopolymer composites enhanced by nano-kaolin material, J. Adv. Ceram. 4 (2015) 245–252.
[73] S. Xu, M.A. Malik, Z. Qi, B. Huang, Q. Li, M. Sarkar, Influence of the PVA fibers and SiO_2 NPs on the structural properties of fly ash based sustainable geopolymer, Construct. Build Mater. 164 (2018) 238–245.
[74] M. Kaur, J. Singh, M. Kaur, Microstructure and strength development of fly ash-based geopolymer mortar: role of nano-metakaolin, Construct. Build Mater. 190 (2018) 672–679.
[75] A. Ravitheja, N.K. Kumar, A study on the effect of nano clay and GGBS on the strength properties of fly ash based geopolymers, Mater. Today Proc. 19 (2019) 273–276.
[76] X. Gao, Q. Yu, H. Brouwers, Characterization of alkali activated slag–fly ash blends containing nano-silica, Construct. Build Mater. 98 (2015) 397–406.
[77] J.M. Their, M. Özakça, Developing geopolymer concrete by using cold-bonded fly ash aggregate, nano-silica, and steel fiber, Construct. Build Mater. 180 (2018) 12–22.
[78] G. Saini, U. Vattipalli, Assessing properties of alkali activated GGBS based self-compacting geopolymer concrete using nano-silica, Case Stud. Constr. Mater. (2020), e00352.
[79] A.A. Ramezanianpour, M.A. Moeini, Mechanical and durability properties of alkali activated slag coating mortars containing nanosilica and silica fume, Construct. Build Mater. 163 (2018) 611–621.
[80] A. Çevik, R. Alzeebaree, G. Humur, A. Niş, M.E. Gülşan, Effect of nano-silica on the chemical durability and mechanical performance of fly ash based geopolymer concrete, Ceram. Int. 44 (2018) 12253–12264.

CHAPTER 8

Nano TiO_2-engineered cementitious materials with self-cleaning properties

Qingxu Jin[a] and Hongyan Ma[b]
[a]Department of Civil and Environmental Engineering, Michigan State University, East Lansing, MI, United States
[b]Department of Civil, Architectural and Environmental Engineering, Missouri University of Science and Technology, Rolla, MO, United States

8.1 Introduction

Since nano-titanium oxide (TiO_2) is relatively inexpensive, safe, and chemically stable and has high compatibility with traditional concrete materials, there is a growing interest in using TiO_2 in the construction industry such as building façades, exterior tiles, roofs, and concrete pavements [1–4]. The initial interest in the use of TiO_2 in construction materials stemmed from its white color, which is due to TiO_2 having no absorption in the visible region [5], and hence for being used as a white pigment in paints, cosmetics, and food coloring since ancient times [6]. In 1956, scholars from Japan started using the photochemical power of TiO_2, activated by ultraviolet (UV) irradiation, in various organic solvents such as alcohols and hydrocarbons [7]. In the early 1970s, a Nature article [8] for the first time reported water photo-electrolysis, also known as the "Honda-Fujishima effect," where a semiconductor of TiO_2 electrode was used to facilitate hydrogen evolution under UV radiation. Since then, the field of TiO_2 photocatalysis has experienced major developments every 10 years for the last several decades. The production of photocatalytic H_2 was reported in the 1980s [5] and hydrophilic TiO_2 films and coatings were developed in the 1990s [9]. The combination of photocatalytic oxidation reactions of adsorbed substances and the photo-induced hydrophilic conversion of TiO_2 itself has led to a broad range of applications, particularly in building materials and environmental applications [5]. During the twenty-first century, TiO_2 has become a practical technology for water treatment of hydroponic systems [5], antibacterial metal applications [10,11], removal of volatile organic compounds (VOCs) [12],

Nanotechnology for Civil Infrastructure
https://doi.org/10.1016/B978-0-12-817832-4.00006-7

air purification through smog abatement [13–15], and surface self-cleaning properties [12,16,17].

In this chapter, we will focus on the "self-cleaning" properties of nano-TiO_2-engineered photocatalytic cementitious materials. We will first explore the self-cleaning mechanism and then introduce two main types of TiO_2-engineered cementitious materials—TiO_2-doped and TiO_2-coated cementitious materials. Since the interaction between TiO_2 and cementitious substrates for each type is different, we will discuss the effects of TiO_2 on the cement for each type, including hydration process, microstructure, strength, surface hydrophilicity, and bond strength between coating and cementitious substrates. Finally, we will present the assessment technique for photocatalytic self-cleaning performance and present its application of methylene blue (MB) photodegradation as an example.

8.2 Self-cleaning mechanism and types of TiO_2-engineered cementitious materials

8.2.1 Self-cleaning mechanism

Self-cleaning properties help create contamination-free surface and in turn maintain the aesthetic characteristics of the concrete urban environment and minimize the cleaning expenses [18,19]. The TiO_2-based self-cleaning mechanism combines photocatalysis and photo-induced superhydrophilicity. Surfaces with contact angles $<90°$ are considered hydrophilic and with contact angles $>90°$ are considered hydrophobic [20–22]. For superhydrophilic surfaces, the contact angle is nearly 0° and water droplets can spread on the surface and form a thin film [20]. When the TiO_2-engineered surfaces are exposed to the sun, the high photocatalytic capability of TiO_2 can be activated (Eqs. 8.1–8.3) in the presence of water, oxygen, and UV or near UV-light, leading to the formation of hydroxyl radicals ($OH^{\bullet}$) and superoxide radicals ($O_2^{\bullet-}$) [17,23].

$$TiO_2 \xrightarrow{hv} TiO_2 + e^- + h^+ \tag{8.1}$$

$$h^+ + H_2O \rightarrow H^+ + OH^{\bullet} \tag{8.2}$$

$$e^- + O_2 \rightarrow O_2^{\bullet-} \tag{8.3}$$

These photocatalytic surfaces are capable of decomposing organic pollutants, including polymers, visible stains, aromatics, dyes, and surfactants [24]. These small particles and greasy deposits such as fatty acids adhere on the building surfaces, causing aesthetic concerns [25]. For example, the carboxylic groups (—COOH) of fatty acids enable them to stick on the building surface by chemically bonding with calcium ions present in concrete [25]. Their long chains link with other hydrophobic molecules known as hydrophobic interactions and perpendicularly extend from the building surface, trapping more atmospheric particles and dust [1]. The decomposition of carboxylic groups based on photocatalytic self-cleaning property is considered a chain reaction. First, the high redox power of photo-induced electron-hole pairs (Eq. 8.4) can decompose the organic binders in the following manner (with palmitic acid as an example).

$$\begin{aligned} n - C_{15}H_{31} - COOH + h^{+} &\rightarrow n - C_{15}H_{31} - COO + H^{+} \\ &\rightarrow n - C_{15}H_{31}^{\bullet} + CO_2 + H^{+} \end{aligned} \tag{8.4}$$

After the release of the first CO_2, the n-$C_{15}H_{31}^{\bullet}$ are oxidized by hydroxyl radicals to an alcohol, and then further oxidized into acid n-$C_{14}H_{29}$ − COOH (Eq. 8.5), which will undergo similar reactions to release the second CO_2. The chain of reaction continues until the acid is completely decomposed to CO_2 and H_2O.

$$\begin{aligned} n - C_{15}H_{31}^{\bullet} + OH^{\bullet} &\rightarrow n - C_{15}H_{31}\,OH + O_2 \\ &\rightarrow n - C_{14}H_{29} - COOH + H_2O \end{aligned} \tag{8.5}$$

8.2.2 Types

In the current building and construction applications, TiO_2 nanoparticle-doped cementitious materials are more widely used than TiO_2-coated cementitious materials due to the concerns of TiO_2 coating might be lost in the field from wear and weathering [26,27]. Research regarding the TiO_2 nanoparticle-blended application has been focused on the effects of cementitious material nanostructure, microstructure, and surface features on photocatalytic performance in the presence of TiO_2 nanoparticles [26,28,29], the effects of mixing procedures and environmental conditions on photocatalytic efficiency [27,30], the effects of TiO_2 nanoparticles on the early hydration and mechanical properties of TiO_2-doped photocatalytic cementitious materials [31,32], and the feasibility of using alternative cementitious materials in combination with TiO_2 nanoparticles [33,34].

Although there are concerns about TiO_2 coating loss with field applications, the use of TiO_2 photocatalytic functional coating on the surface of cementitious material is attracting a growing attention for two major reasons. First, because the photocatalytic reactions are surface phenomena

[35], it would be more effective and economically efficient to apply TiO_2-modified functional coatings to the surface of cementitious materials [35]. Second, the TiO_2-coated cementitious materials are considered to have less embodied energy than TiO_2-doped ones [36]. A life cycle analysis carried out by Jayapalan et al. [36] comparing a TiO_2-based cement paste with plain cement paste showed that TiO_2-based cement nearly doubles the material's energy embodiment, even when only 5% by weight TiO_2 is blended in the cement paste. Therefore, TiO_2 coating, using as the functional component on the surface of cementitious substrates as a strategic application, could minimize the TiO_2 dosage, and hence its energy embodiment.

Moreover, to address the durability concerns of a TiO_2 coating and reduce the production cost, previous studies have proposed using titania-silica (TiO_2-SiO_2) composites as a photocatalytic coating since SiO_2 has higher mechanical and thermal stability than TiO_2 [37–39]. Compared to TiO_2, the production of SiO_2 is less expensive [37], and the introduction of SiO_2 also increases the photocatalytic efficiency because its large surface area helps adsorb pollutants and intermediates for longer periods [37]. SiO_2 can also alter the hydrophilicity of TiO_2-SiO_2 coatings [40] and in turn, help address the durability issues of cementitious substrates due to the penetration of water and aggressive chemicals when exposed to atmospheric conditions of relative humidity >50% [41]. Although the addition of SiO_2 improves the durability of the coating itself, the bond between the coating and the surface of cementitious materials must be carefully examined, and the durability of the bond must be ensured to avoid compromising the durability performance of the coated photocatalytic cementitious materials.

8.3 Effect of TiO_2 on cementitious substrates

For TiO_2 nanoparticle-doped cementitious materials, the inclusion of nanoparticles could affect the early hydration, microstructure, and strength of the materials. For TiO_2-coated cementitious materials, the main influences of TiO_2 on materials are surface hydrophilicity and bond strength between coating and cementitious substrates. In this section, we will discuss the effects of TiO_2 on the cement for each type of TiO_2-engineered cementitious materials.

8.3.1 TiO_2 nanoparticle-doped cementitious materials

8.3.1.1 Hydration and strength

Two different effects of TiO_2 inclusion on the hydration of cement have been reported previously: (i) the pozzolanic activity of TiO_2 in promoting the early-age hydration of cement [42,43]; and (ii) being considered as an inert filler, the effect of TiO_2 on the promotion of early-age hydration is likely due to enhanced nucleation and growth of C—S—H surrounding TiO_2 particles [31,36]. Although an agreement has not been reached yet, the effects of TiO_2 additions on cement hydration depend on the cement matrix, water-to-cement ratio (w/c), TiO_2 nanoparticle content, TiO_2 nanoparticle size, and dispersion degree. For example, Jayapalan et al. [36] have observed that the rate of heat release and total heat release of TiO_2 nanoparticle-doped cementitious materials increases with the increased content, as shown in Fig. 8.1. For supporting the nucleation and growth effects, the authors claimed that the nucleation effect dominates the dilution effect since higher TiO_2 replacement (wt% of cement) increases the rate of early reaction. More importantly, the more well-dispersed nanoparticles, which according to Jayapalan et al. are described in terms of particle and agglomerate sizes, the higher the peak rate of heat release and the total heat release [36]. In addition to cement, Lee and Kurtis [44] carried out similar hydration studies on alite and belite, two of the main clinker minerals of Portland cement. Again, the early-age hydration of alite and belite is accelerated by a TiO_2 inclusion similar to cement, and the authors claimed that the acceleration is also likely due to the nucleation and growth effects.

Lee [32] has measured the compressive strength evolution of cement paste containing 0% (control), 5%, and 10% TiO_2 at 1, 3, 7, 14, and 28 days of age at a w/c of 0.5, as shown in Fig. 8.2. The compressive strength increased until 14 days and reached a plateau. The author also performed a single factor analysis of variance test (ANOVA test) with data at 28 days of curing and concluded that the compressive strength for each case at 28 days of curing is statistically the same, even though the samples with 5% and 10% TiO_2 replacements contain less cement compared to the control samples. This result indicates that the acceleration of early-age hydration induced by TiO_2 could help compensate the potential low strength that might be caused by low cement content from TiO_2-doped cementitious materials.

8.3.1.2 TiO_2 dispersion and microstructure

One of the key factors to ensure a uniform photocatalytic performance is the uniform distribution of TiO_2 nanoparticles within the hydrated cement.

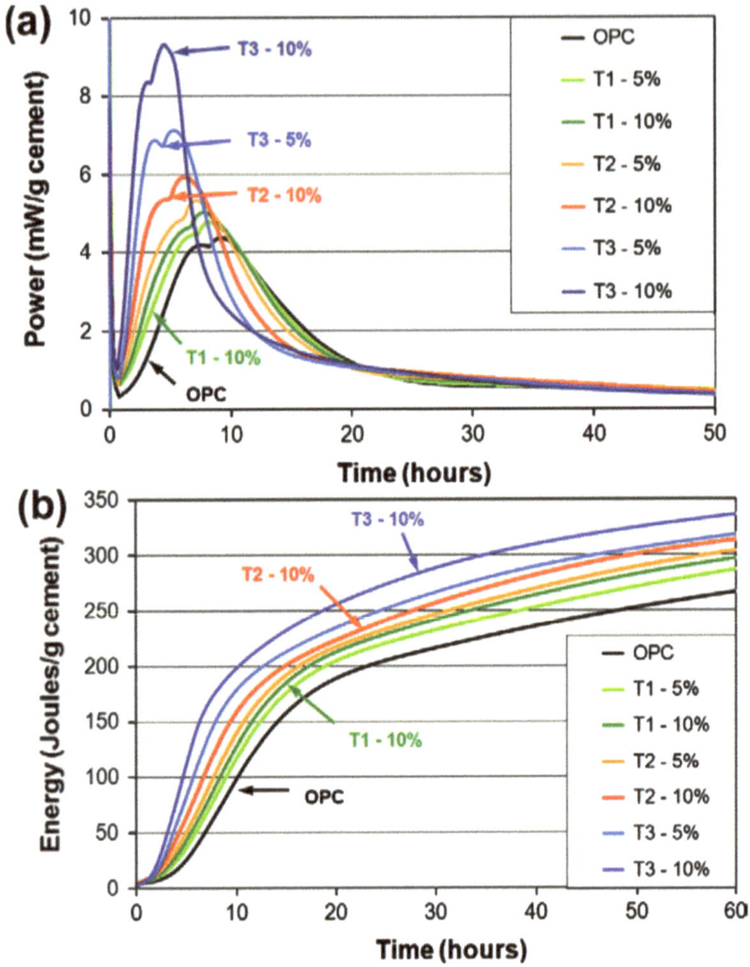

Fig. 8.1 (A) Rate of heat release and (B) total heat release of TiO_2-doped cement samples with different TiO_2 content. T1 has a particle size of 20–30nm and agglomerate size of 1.5μm, T2 has a particle size of 15–25nm agglomerate size of 1.2μm, and T3 has a particle size of 21nm and agglomerate size of 0.5μm [36].

Studies have shown that TiO_2 nanoparticles exhibit a tendency toward agglomeration [45], which could affect the photocatalytic performance. To enhance the particle dispersibility in an aqueous environment, chemical surface treatment using phosphorous and potassium has been performed on the surface of these nanoparticles [45]. Such treatment has also been used in commercial products [46]. In addition to agglomeration, the tendency of TiO_2 nanoparticles binding with any particular hydrated phases could also potentially affect the photocatalytic efficiency. Therefore, to examine the

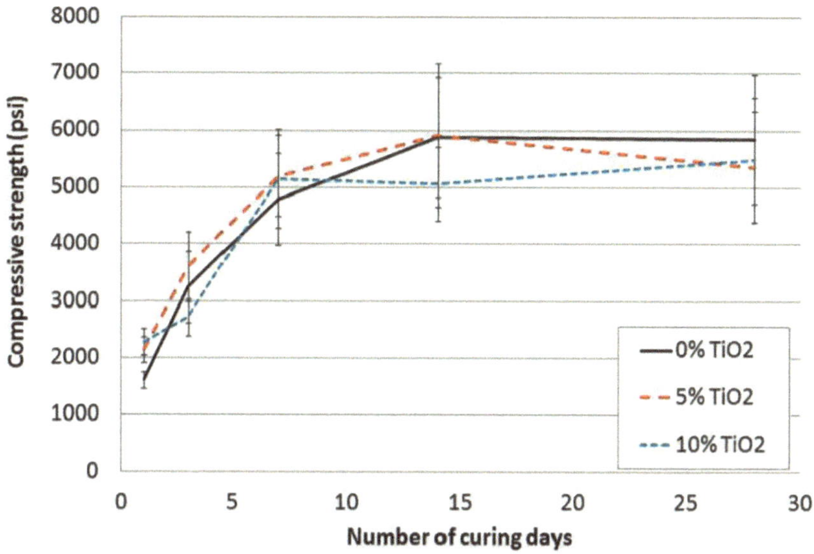

Fig. 8.2 Compressive strength of 0%, 5%, and 10% TiO_2-doped cement pastes at a w/c of 0.5. *(Reproduced from B.Y. Lee, Effect of Titanium Dioxide Nanoparticles on Early Age and Long Term Properties of Cementitious Materials, 2012.)*

TiO_2 nanoparticle dispersion and the corresponding microstructure, Jin et al. [34] recently carried out a scanning electron microscopy (SEM) analysis and observed that the commercial TiO_2 nanoparticles with phosphorous surface treatment are uniformly distributed within calcium aluminate cement (CAC) and exhibit no preference to bind with any particular hydrated phases (Fig. 8.3). However, the image also shows that these TiO_2 particles are in the size range of 1–2 μm and could be the

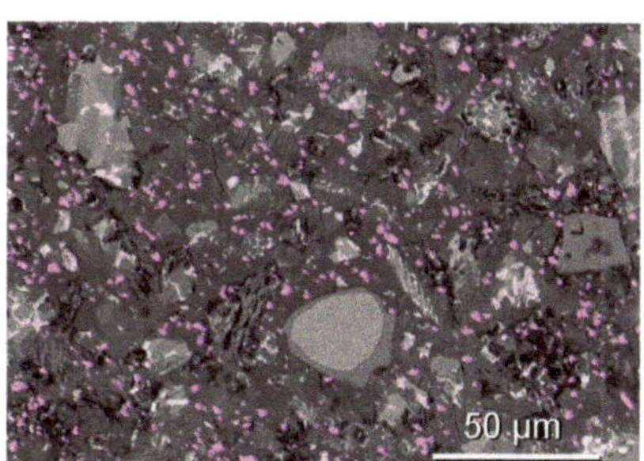

Fig. 8.3 SEM analysis of TiO_2-doped CAC samples, with TiO_2 highlighted [34].

agglomerations of the nanoparticles. Similar observations have also been found in TiO_2-doped ordinary Portland cement (OPC)-based materials [46].

To understand the effect of TiO_2 nanoparticles on the material's microstructural properties, Jin et al. [34,47] prepared the granular cementitious samples with the size of 0.6–1 mm and measured their specific surface areas (SSA) using N_2 adsorption and desorption isotherms. The results show that the samples with TiO_2 nanoparticles have higher SSA and larger amount of small pores (<10 nm) compared to control samples without TiO_2. These effects should be mainly attributed to the inclusion of TiO_2 nanoparticles, which intrinsically inherit high surface area [34,47].

8.3.2 TiO_2-coated cementitious materials

8.3.2.1 Hydrophilicity

For TiO_2-coated cementitious materials, the TiO_2 coating has little effect on the hydration and strength since the coating is usually applied on the surface after 28 days of curing and hydration of cementitious substrate [48,49]. However, the coating could alter the hydrophilicity of the photocatalytic cementitious material surfaces. As mentioned in Section 1.2, hydrophilicity could affect the photocatalytic performance of TiO_2-coated cementitious materials and can be characterized by the wettability test, which essentially examines the interactions between the sample surfaces and water. The cementitious materials are usually hydrophilic with a contact angle of 10°–15° [48] and the TiO_2 coating could help the surface achieve superhydrophilic with a contact angle of 0° [48]. However, by including a hydrophobic SiO_2 layer in between TiO_2 coating and cementitious substrate, the surface can be altered to hydrophobic or superhydrophobic with contact angles larger than 100° [22,48]. The ability to change the surface hydrophilicity and hydrophobicity lends the TiO_2-SiO_2 nanoparticle-modified coatings to broad applications in addition to photocatalytic performance. For example, in repair applications, hydrophobic coatings are more favorable since they could help prevent water from penetrating into concrete, potentially avoiding future repairs in the same area.

8.3.2.2 Bond strength

Another important parameter of TiO_2-coated cementitious materials is the bond strength between the functional TiO_2 coating and cementitious substrate, which is the largest concern of applying TiO_2-coated cementitious materials in infrastructure such as pavement surface that is directly exposed to the traffic loading. It has been demonstrated that the inclusion of a SiO_2 layer could affect the bond strength between coating and cementitious materials [48], which could be evaluated by adhesion tests (ASTM D3359). The results of adhesion tests are shown in Fig. 8.4 [48]. The PB and C4B samples are plain OPC and OPC pastes, respectively. The samples coated with only TiO_2 layer are denoted as PT and C4T, the samples with only hydrophobic SiO_2 layer are denoted as PS1 and C4S1, and the samples with both TiO_2 and SiO_2 are denoted PST1 and C4ST1 for OPC and CAC, respectively. The results show that in samples with only TiO_2 layer, the coatings can be easily peeled off during the cut without damaging the surfaces of cementitious substrates, indicating a weak bond between these coatings and cementitious substrates. In contrast, the surfaces of samples coated with a hydrophobic SiO_2 layer are irregular and roughened after the cut. This result indicates a strong bonding between coatings and cementitious substrates as the damage occurs in the cementitious substrate when peeling off the coatings. This strong bond performance appears to be related to the hydrophobicity of the SiO_2 layer. For coated concrete infrastructure where a strong bond between coating and cementitious substrate is desirable, an addition of hydrophobic SiO_2 layer is recommended for improving the bond strength.

8.4 Self-cleaning performance and applications

8.4.1 Assessment techniques

To determine the self-cleaning activity of the surface of the TiO_2-engineered cementitious materials, ISO standard: 10678:2010—"Determination of photocatalytic activity of surface in aqueous medium by degradation of methylene blue"—has been widely used [50]. Although this standard is based on water remediation, it still provides useful guidance to measure the ability of a photocatalytic coated surface to clean the deposited organic species by observing the destruction of a pollutant. To demonstrate this effect, a dye, specifically MB dye, is used as a test pollutant since only a small amount of material is needed to show a striking color change as the dye is photo-oxidized. The complete photodecomposition of MB dye ($C_{16}H_{18}N_3SCl$) by titania films follows the reaction below [51]:

$$C_{16}H_{18}N_3SCl + 25\frac{1}{2}O_2 \xrightarrow{TiO_2,\, hv \geq 3.2\ eV} HCl + H_2SO_4 + HNO_3 + 16CO_2 + 6H_2O \tag{8.6}$$

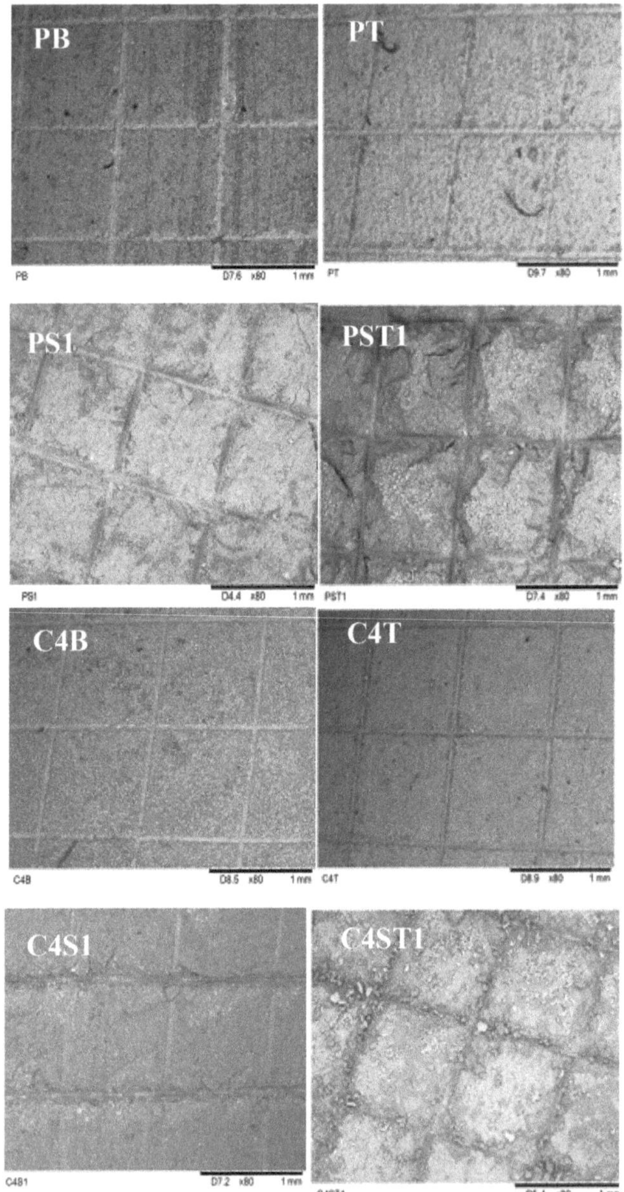

Fig. 8.4 Adhesion tests for examining bond strength between TiO_2 coating and cementitious substrates. *(Modified from ACI special publication Q. Jin, M. Faraldos, A. Bahamonde, B.H. Zaribaf, K.E. Kurtis, Titania and silica nanoparticle-modified coatings for cementitious materials, ACI Spec. Publ. 335 (2019) 97–111.)*

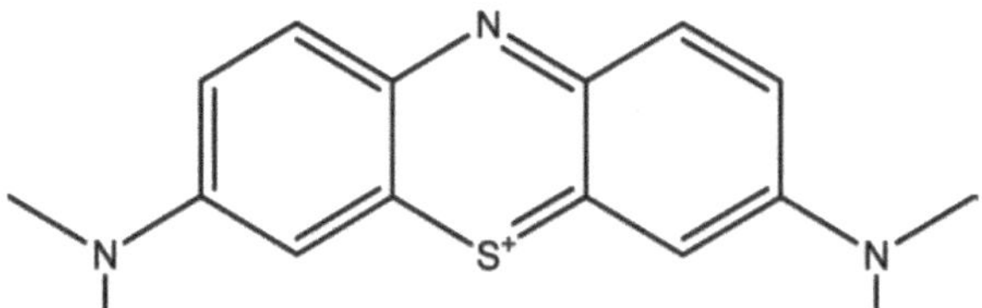

MB

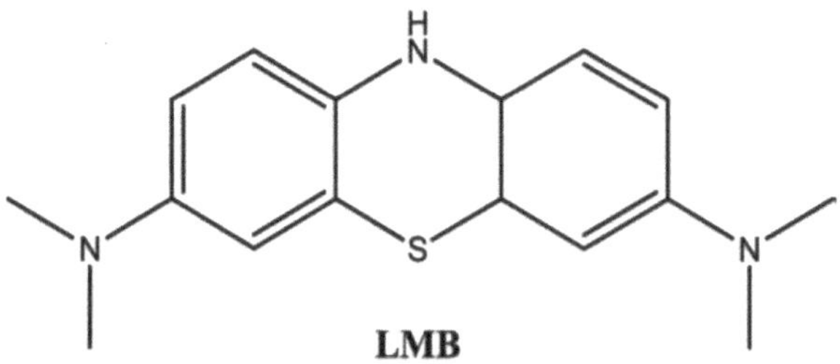

LMB

Fig. 8.5 Structure of MB dye and its colorless leuco-MB (LMB) [52].

The common chemical structure of MB is shown in Fig. 8.5 [52], together with its colorless and reduced form of leuco-MB (LMB). The popularity of using MB dye for the test is due to its large molar absorptivity, nontoxic nature, high solubility in water, stability in an aqueous solution, and its ongoing applications and availability for textile, leather, and paper dye [52].

The standardized test [50] uses a plate sample of $10\,cm^2$ with an active coating and a glass cylinder with a diameter of about 4 cm is fixed on top of the sample. An MB solution containing 2×10^{-5} mmol of MB is prepared and stored in the dark for 12 h before the test, during which if the concentration of MB reaches lower than 1×10^{-5} mmol preparation must be repeated using a fresh conditioning solution. After that, 35 mL of solution is placed in the cylinder, which is then covered with a UV transparent glass pane, such as borosilicate. The system is then under the radiation of a UV-light with an energy density of $1.0\,mW/m^2$ and a narrow-band emitter such as a blacklight blue (BLB) lamp should be used. The concentration of the MB solution is measured by a spectrophotometer at a wavelength of 664 nm as a function of irradiation time. The irradiation process is carried out for 3 h, or until whenever the solution is decolorized, whichever happens first. The reaction temperature should be $23 \pm 2°C$. A blank control group should also be tested under a dark environment. The photocatalytic efficiency of the system can be calculated based on the relationship between the MB concentration and irradiation time. For testing the self-cleaning properties of photocatalytic cementitious materials, previous studies have applied MB dye drops on the coated surfaces and a similar measurement of discoloration is followed as previously mentioned [53].

The biggest advantage of the method presented above is its simplicity. However, this simplicity is based on several underlying assumptions, which could cause potential errors and increase uncertainties. The main assumptions that could be problematic have been identified by Mills [52], who also provided suggestions for improving such concerns. The ones that are the most relevant to self-cleaning cementitious materials are listed here: (i) the purity of MB dye: studies have shown significant variations in an impure MB source [54], and Mills [52] suggested to use a known source of MB dye and self-validate its purity before using it; (ii) pH: a pH, whose value is below the zero charged substrate, would be ideal for the relevant test. In the case of TiO_2, the pH value of the zero charger should be 6.6, therefore, Mills [52] suggested the initial pH of the MB dye should be in the range of 5.5–6.0 for the test; (iii) light source: the standard recommends a narrow-band light source in the range of 320–400 nm. Mills [52] argued that such a loosely defined light source could cause significant variations and suggested that the BLB UV-light with a europium-doped strontium fluoroborate or borate phosphor should be used. Despite these concerns, previous studies [53] have demonstrated that the application of such tests in photocatalytic cementitious materials can still provide useful insights, in particular the correlation between MB photodegradation and photocatalytic self-cleaning behavior.

8.4.2 Methylene blue photodegradation

The results of MB photodegradation could provide two sets of information: (1) the indication of hydrophilicity of the coated surface; i.e., the greater hydrophilicity of the surface the larger spread area of MB solution per the same volume, and (2) the lightening of MB solution is the indication of the photocatalytic degradation of MB solution, the mechanism of which is presented in the previous section. The self-cleaning efficiency of TiO_2-doped cementitious materials has rarely been tested, since the doping is more relevant to applications for air/water cleaning while self-cleaning is a surface phenomenon [55]. Fig. 8.6 shows the MB photodegradation of TiO_2-coated OPC pastes and CAC pastes along with the radiation time [48]. It was observed that cementitious materials with a TiO_2 layer (with sample ID of PT, PST1, PST2, C4T, C4ST1, and C4ST2 in Fig. 8.6) show

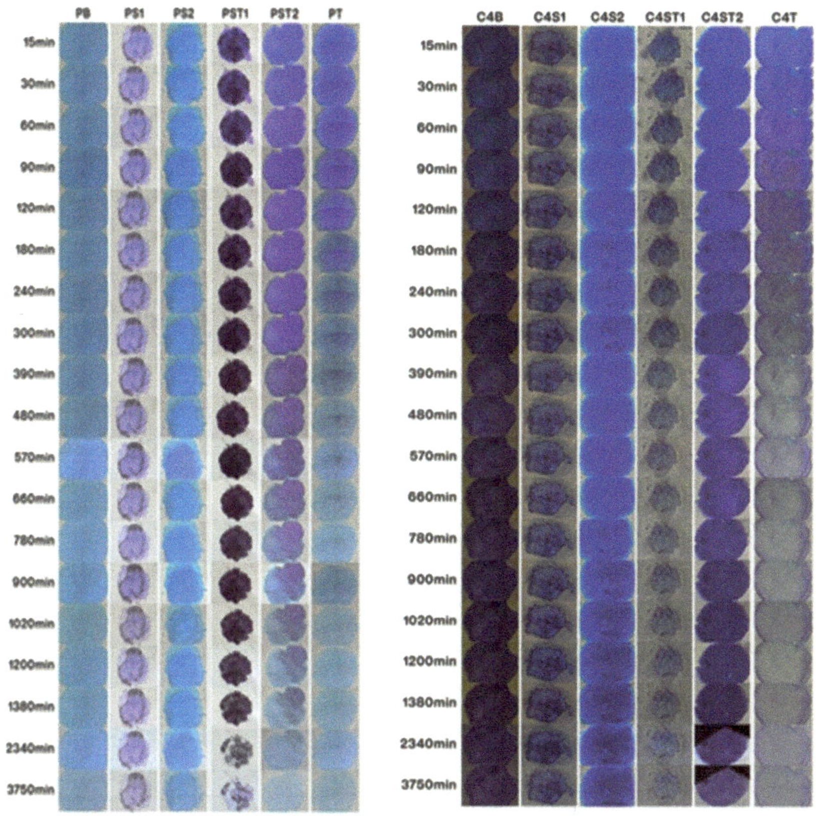

Fig. 8.6 MB dye degradation of TiO_2-coated Portland cement pastes *(left)* and CAC pastes *(right)*.

a significant MB photodegradation (blue color fading) regardless of the inclusion of SiO_2 layer or not, while the ones without TiO_2 (with sample ID of PB, PS1, PS2, C4B, C4S1, and C4S2 in Fig. 8.6) show insignificant MB photodegradation even with SiO_2 layer. This comparison suggests that the coatings containing TiO_2 promote the photocatalytic activity and hence increase the rate of MB photodegradation. Fig. 8.7 plots the efficiency of MB photodegradation along with the radiation time. The time to achieve 50% MB dye bleaching ($t_{1/2}$) is indicated by the red dashed line. The sample with only TiO_2 layers presents a faster 50% MB dye bleaching (lower value of $t_{1/2}$) than samples with both TiO_2 and SiO_2 layers, indicating that the inclusion of silica layers may affect the interaction between coatings and cementitious substrates, hence, reduces the MB photodegradation efficiency. Jin et al. [48] have also examined the microstructure of the two

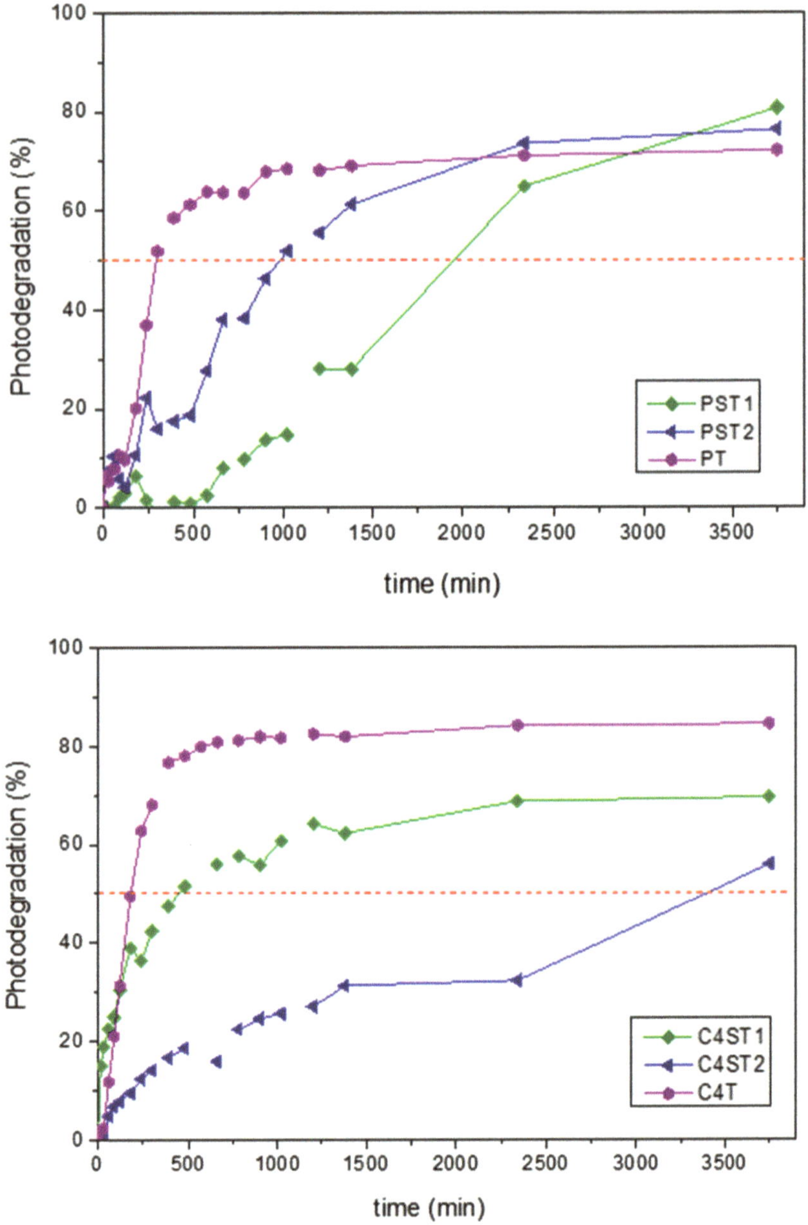

Fig. 8.7 MB dye degradation efficiency of TiO_2-coated cementitious composite—Portland cement (PT, PST1, and PST2) and CAC (C4T, C4ST1, and C4ST2). *(Reproduced from ACI special publication Q. Jin, M. Faraldos, A. Bahamonde, B.H. Zaribaf, K.E. Kurtis, Titania and silica nanoparticle-modified coatings for cementitious materials, ACI Spec. Publ. 335 (2019) 97–111.)*

cementitious substrates (i.e., OPC vs CAC) and found a significant difference between these cementitious materials in terms of porosity and pore distribution. However, their examination revealed little difference between coated and uncoated samples, suggesting the coatings have an insignificant effect on the microstructures of these two TiO_2-coated cementitious materials. Moreover, the authors also found that the differences in microstructure seem to play an insignificant role in the MB photodegradation compared to the differences in the coating composition, which indicates the MB photodegradation largely depend on the photocatalytic performance of the coatings themselves [48].

8.5 Conclusion and future perspectives

The effects of TiO_2 nanoparticles on microstructure and properties of TiO_2-doped and TiO_2-coated cementitious materials, and the resultant self-cleaning property (as well as the mechanisms) are discussed in this chapter. The self-cleaning capacity of these materials can be attributed to TiO_2-induced photocatalysis and superhydrophilicity: the former effect leads to continuous photocatalytic decomposition of organic pollutants, and the latter effect results in easier spreading of pollutants and facilitates photocatalytic reactions.

In TiO_2-doped cementitious materials, the incorporated TiO_2 nanoparticles manifest a filler effect, diluting cement and accelerating the early hydration of cement. The filler effect typically results in an increased early-age strength but no obvious effect on late age strength (e.g., at the age of 28 days). The detailed effects of TiO_2 on cementitious materials depend on how well the nanoparticles are dispersed.

In TiO_2-coated cementitious materials, the TiO_2 coating improves the surface hydrophilicity, but the bonding between the coating and cement substrate can be weak. Inserting a SiO_2 layer between the TiO_2 coating and the substrate can make the coating hydrophobic and improve largely the interfacial bond strength.

Since self-cleaning is a surface phenomenon, TiO_2 coating is a more feasible strategy as compared to doping. The efficiency of self-cleaning of TiO_2-coated cement-based materials, as demonstrated using photodegradation of MB, has been well documented.

As the self-cleaning capacity of TiO_2-functionalized cement-based materials is normally stimulated by UV or near UV-light—which limits the application of self-cleaning—it will be attractive in future studies to enable visible-light-responsive photocatalysis. In addition, workability (ease of application of photocatalytic coating) and strength of coating are also practical issues to be addressed in the future.

References

[1] J. Chen, C.s. Poon, Photocatalytic construction and building materials: from fundamentals to applications, Build. Environ. 44 (2009), https://doi.org/10.1016/j.buildenv.2009.01.002.

[2] M.M. Hassan, H. Dylla, L.N. Mohammad, T. Rupnow, Evaluation of the durability of titanium dioxide photocatalyst coating for concrete pavement, Constr. Build. Mater. 24 (2010) 1456–1461, https://doi.org/10.1016/j.conbuildmat.2010.01.009.

[3] C.J. Cros, A.L. Terpeluk, L.E. Burris, N.E. Crain, R.L. Corsi, M.C.G. Juenger, Effect of weathering and traffic exposure on removal of nitrogen oxides by photocatalytic coatings on roadside concrete structures, Mater. Struct. 48 (2014), https://doi.org/10.1617/s11527-014-0388-2.

[4] H.J.H. Brouwers, M.M. Ballari, M. Hunger, G. Hu, NOx photocatalytic degradation employing concrete pavement containing titanium dioxide, Appl. Catal. B Environ. 95 (2010), https://doi.org/10.1016/j.apcatb.2010.01.002.

[5] K. Hashimoto, H. Irie, A. Fujishima, TiO_2 photocatalysis: a historical overview and future prospects, Jpn. J. Appl. Phys. 44 (2005) 8269–8285. https://www.jsap.or.jp/jsapi/Pdf/Number14/04_JJAP-IRP.pdf. (Accessed 22 September 2018).

[6] M.J. Gázquez, J.P. Bolívar, R. Garcia-Tenorio, F. Vaca, A review of the production cycle of titanium dioxide pigment, Mater. Sci. Appl. 05 (2014) 441–458, https://doi.org/10.4236/msa.2014.57048.

[7] F. Kato, S. Mashio, Autooxidation by TiO_2 as a photocatalyst, Annu. Meet Chem. Soc. Jpn. 223 (1956).

[8] A. Fujishima, K. Honda, Electrochemical photolysis of water at a semiconductor electrode, Nature 238 (1972), https://doi.org/10.1038/238037a0.

[9] R. Wang, K. Hashimoto, A. Fujishima, M. Chikuni, E. Kojima, A. Kitamura, M. Shimohigoshi, T. Watanabe, Light-induced amphiphilic surfaces, Nature 388 (1997) 431–432, https://doi.org/10.1038/41233.

[10] D.J. Giannantonio, J.C. Kurth, K.E. Kurtis, P.A. Sobecky, Effects of concrete properties and nutrients on fungal colonization and fouling, Int. Biodeterior. Biodegradation 63 (2008) 252–259, https://doi.org/10.1016/j.ibiod.2008.10.002.

[11] A. Maury Ramirez, N. De Belie, Evaluation of the Algaecide Activity of Titanium Dioxide on Autoclaved Aerated Concrete, [STI, Science & Technology Integration], 2009. https://biblio.ugent.be/publication/878143. (Accessed 22 September 2018).

[12] S. Wang, H.M. Ang, M.O. Tade, Volatile organic compounds in indoor environment and photocatalytic oxidation: state of the art, Environ. Int. 33 (2007) 694–705, https://doi.org/10.1016/j.envint.2007.02.011.

[13] J. Chen, C.S. Poon, Photocatalytic activity of titanium dioxide modified concrete materials - influence of utilizing recycled glass cullets as aggregates, J. Environ. Manag. 90 (2009), https://doi.org/10.1016/j.jenvman.2009.05.029.

[14] Q. Jin, E.M. Saad, W. Zhang, Y. Tang, K.E. Kurtis, Quantification of NO_x uptake in plain and TiO_2−doped cementitious materials, Cem. Concr. Res. (2019), https://doi.org/10.1016/j.cemconres.2019.05.010.

[15] J. Chen, S.c. Kou, C.s. Poon, Photocatalytic cement-based materials: comparison of nitrogen oxides and toluene removal potentials and evaluation of self-cleaning performance, Build. Environ. 46 (2011), https://doi.org/10.1016/j.buildenv.2011.03.004.

[16] L. Cassar, Photocatalysis of cementitious materials: clean buildings and clean air, MRS Bull. 29 (2004) 328–331, https://doi.org/10.1557/mrs2004.99.
[17] A. Folli, C. Pade, T.B. Hansen, T. De Marco, D.E. MacPhee, TiO_2 photocatalysis in cementitious systems: insights into self-cleaning and depollution chemistry, Cem. Concr. Res. 42 (2012), https://doi.org/10.1016/j.cemconres.2011.12.001.
[18] L. Pinho, M.J. Mosquera, Photocatalytic activity of TiO_2–SiO_2 nanocomposites applied to buildings: influence of particle size and loading, Appl. Catal. B Environ. 134–135 (2013), https://doi.org/10.1016/j.apcatb.2013.01.021.
[19] A. Fujishima, X. Zhang, D.A. Tryk, TiO_2 photocatalysis and related surface phenomena, Surf. Sci. Rep. 63 (2008) 515–582, https://doi.org/10.1016/J.SURFREP.2008.10.001.
[20] J.T. Simpson, S.R. Hunter, T. Aytug, Superhydrophobic materials and coatings: a review, Rep. Prog. Phys. 78 (2015), https://doi.org/10.1088/0034-4885/78/8/086501.
[21] A. Marmur, C. Della Volpe, S. Siboni, A. Amirfazli, J.W. Drelich, Contact angles and wettability: towards common and accurate terminology, Surf. Innov. 5 (2017) 3–8, https://doi.org/10.1680/jsuin.17.00002.
[22] I. Flores-Vivian, V. Hejazi, M.I. Kozhukhova, M. Nosonovsky, K. Sobolev, Self-assembling particle-siloxane coatings for superhydrophobic concrete, ACS Appl. Mater. Interfaces 5 (2013) 13284–13294, https://doi.org/10.1021/am404272v.
[23] M.M. Ballari, M. Hunger, G. Hüsken, H.J.H. Brouwers, Modelling and experimental study of the NOx photocatalytic degradation employing concrete pavement with titanium dioxide, Catal. Today 151 (2010), https://doi.org/10.1016/j.cattod.2010.03.042.
[24] I.P. Parkin, R.G. Palgrave, Self-cleaning coatings, J. Mater. Chem. 15 (2005) 1689–1695.
[25] L. Peruchon, E. Puzenat, A. Girard-Egrot, L. Blum, J.M. Herrmann, C. Guillard, Characterization of self-cleaning glasses using Langmuir–Blodgett technique to control thickness of stearic acid multilayers: importance of spectral emission to define standard test, J. Photochem. Photobiol. A Chem. 197 (2008) 170–176, https://doi.org/10.1016/J.JPHOTOCHEM.2007.12.033.
[26] J. Chen, C.S. Poon, Photocatalytic cementitious materials: influence of the microstructure of cement paste on photocatalytic pollution degradation, Environ. Sci. Technol. 43 (2009), https://doi.org/10.1021/es902359s.
[27] M. Guo, C. Poon, Photocatalytic NO removal of concrete surface layers intermixed with, Build. Environ. 70 (2013), https://doi.org/10.1016/j.buildenv.2013.08.017.
[28] P. Alina, A. Janina, L. Bogdan, Application of Titanium Dioxide in Cement and Concrete Technology, vol. 687, 2016, https://doi.org/10.4028/www.scientific.net/KEM.687.243.
[29] A.M. Ramirez, K. Demeestere, N. De Belie, T. Mantyla, E. Levanen, Titanium dioxide coated cementitious materials for air purifying purposes: preparation, characterization and toluene removal potential, Build. Environ. 45 (2010), https://doi.org/10.1016/j.buildenv.2009.09.003.
[30] A. Yousefi, A. Allahverdi, P. Hejazi, Effective dispersion of nano-TiO_2 powder for enhancement of photocatalytic properties in cement mixes, Constr. Build. Mater. 41 (2013), https://doi.org/10.1016/j.conbuildmat.2012.11.057.
[31] A.R. Jayapalan, B.Y. Lee, S.M. Fredrich, K.E. Kurtis, Influence of additions of anatase TiO_2 nanoparticles on early-age properties of cement-based materials, Transp. Res. Rec. 2 (2010) 41–46, https://doi.org/10.3141/2141-08.
[32] B.Y. Lee, Effect of Titanium Dioxide Nanoparticles on Early Age and Long Term Properties of Cementitious Materials, 2012.

[33] M. Pérez-nicolás, J. Balbuena, M. Cruz-yusta, L. Sánchez, I. Navarro-blasco, J.M. Fernández, J.I. Alvarez, Photocatalytic NOx abatement by calcium aluminate cements modified with TiO_2: improved NO_2 conversion, Cem. Concr. Res. 70 (2015), https://doi.org/10.1016/j.cemconres.2015.01.011.
[34] Q. Jin, S.L. Hordern, Y. Tang, K.E. Kurtis, NOx sequestration by calcium aluminate cementitious materials, Cem. Concr. Res. 142 (2021), 106381, https://doi.org/10.1016/j.cemconres.2021.106381.
[35] A. Folli, S.B. Campbell, J.A. Anderson, D.E. MacPhee, Role of TiO_2 surface hydration on NO oxidation photo-activity, J. Photochem. Photobiol. A Chem. 220 (2011), https://doi.org/10.1016/j.jphotochem.2011.03.017.
[36] A.R. Jayapalan, B.Y. Lee, K.E. Kurtis, Can nanotechnology be “green”? Comparing efficacy of nano and microparticles in cementitious materials, Cem. Concr. Compos. 36 (2013), https://doi.org/10.1016/j.cemconcomp.2012.11.002.
[37] Y. Hendrix, A. Lazaro, Q. Yu, J. Brouwers, Titania-silica composites: a review on the photocatalytic activity and synthesis methods, World J. Nano Sci. Eng. 05 (2015), https://doi.org/10.4236/wjnse.2015.54018.
[38] M. Montes, F.P. Getton, M.S.W. Vong, P.A. Sermon, Titania on silica. A comparison of sol-gel routes and traditional methods, J. Sol-Gel Sci. Technol. 8 (1997) 131–137, https://doi.org/10.1023/A:1026440213534.
[39] C. Huang, H. Bai, Y. Huang, S. Liu, S. Yen, Y. Tseng, Synthesis of neutral/hydrosol and its application as antireflective self-cleaning thin film, Int. J. Photoenergy 2012 (2012) 1–8, https://doi.org/10.1155/2012/620764.
[40] S. Banerjee, D.D. Dionysiou, S.C. Pillai, Self-cleaning applications of TiO_2 by photo-induced hydrophilicity and photocatalysis, Appl. Catal. B Environ. 176–177 (2015) 396–428, https://doi.org/10.1016/j.apcatb.2015.03.058.
[41] R.W.B.P.K. Mehta, Building durable structures in the 21st century, Concr. Int. 23 (2001) 57–63. http://citeseerx.ist.psu.edu/viewdoc/download?doi=10.1.1.475.5570&rep=rep1&type=pdf.
[42] L.Y. Yang, Z.J. Jia, Y.M. Zhang, J.G. Dai, Effects of nano-TiO_2 on strength, shrinkage and microstructure of alkali activated slag pastes, Cem. Concr. Compos. 57 (2015) 1–7, https://doi.org/10.1016/j.cemconcomp.2014.11.009.
[43] Z. Li, B. Han, X. Yu, S. Dong, L. Zhang, X. Dong, J. Ou, Effect of nano-titanium dioxide on mechanical and electrical properties and microstructure of reactive powder concrete, Mater. Res. Express 4 (2017), 095008, https://doi.org/10.1088/2053-1591/aa87db.
[44] B.Y. Lee, K.E. Kurtis, Influence of TiO_2 nanoparticles on early C3S hydration, J. Am. Ceram. Soc. 93 (2010) 3399–3405, https://doi.org/10.1111/j.1551-2916.2010.03868.x.
[45] A. Folli, I. Pochard, A. Nonat, U.H. Jakobsen, A.M. Shepherd, D.E. Macphee, Engineering photocatalytic cements: understanding TiO_2 surface chemistry to control and modulate photocatalytic performances, J. Am. Ceram. Soc. 93 (2010) 3360–3369, https://doi.org/10.1111/j.1551-2916.2010.03838.x.
[46] Q. Jin, Fundamental understanding of NO_x sequestration capacity and pathways, in: Nano-TiO_2 Engineered Cementitious Materials, Georgia Institute of Technology, 2019.
[47] Q. Jin, E.M. Saad, W. Zhang, Y. Tang, K.E. Kurtis, Quantification of NO_x uptake in plain and TiO_2-doped cementitious materials, Cem. Concr. Res. 122 (2019) 251–256, https://doi.org/10.1016/J.CEMCONRES.2019.05.010.
[48] Q. Jin, M. Faraldos, A. Bahamonde, B.H. Zaribaf, K.E. Kurtis, Titania and silica nanoparticle-modified coatings for cementitious materials, ACI Spec. Publ. 335 (2019) 97–111.

[49] L. Yang, A. Hakki, F. Wang, D.E. Macphee, Photocatalyst efficiencies in concrete technology: the effect of photocatalyst placement, Appl. Catal. B Environ. 222 (2018) 200–208, https://doi.org/10.1016/j.apcatb.2017.10.013.
[50] ISO/TC 206 Fine Ceramics, ISO 10678:2010 Fine Ceramics (Advanced Ceramics, Advanced Technical Ceramics) — Determination of Photocatalytic Activity of Surfaces in an Aqueous Medium by Degradation of Methylene Blue, 2010.
[51] R.W. Matthews, Photocatalytic oxidation and adsorption of methylene blue on thin films of near-ultraviolet-illuminated TiO_2, J. Chem. Soc. Faraday Trans. 1 Phys. Chem. Condens. Phases 85 (1989) 1291–1302, https://doi.org/10.1039/F19898501291.
[52] A. Mills, An overview of the methylene blue ISO test for assessing the activities of photocatalytic films, Appl. Catal. B Environ. 128 (2012) 144–149, https://doi.org/10.1016/J.APCATB.2012.01.019.
[53] C. Mendoza, A. Valle, M. Castellote, A. Bahamonde, M. Faraldos, TiO_2 and TiO_2-SiO_2 coated cement: comparison of mechanic and photocatalytic properties, Appl. Catal. B Environ. 178 (2015) 155–164, https://doi.org/10.1016/j.apcatb.2014.09.079.
[54] S. Mowry, P.J. Ogren, Kinetics of methylene blue reduction by ascorbic acid, J. Chem. Educ. 76 (1999) 970, https://doi.org/10.1021/ed076p970.
[55] J. Chen, S. Kou, C. Poon, Photocatalytic cement-based materials: comparison of nitrogen oxides and toluene removal potentials and evaluation of self-cleaning performance, Build. Environ. 49 (2011) 1827–1833, https://doi.org/10.1016/j.buildenv.2011.03.004.

Index

Note: Page numbers followed by *f* indicate figures and *t* indicate tables.

G

FSC
www.fsc.org
MIX
Papier | Fördert
gute Waldnutzung
FSC® C083411

Kolibri 360 GmbH
Ferdinand-Jühlke-Straße 7
99095 Erfurt, Deutschland
produktsicherheit@kolibri360.de